The Physics of Chance

The Physics of Chance
From Blaise Pascal to Niels Bohr

Charles Ruhla
Professor, Claude-Bernard University (Lyon-1)

Translated from the French (*La physique du hasard: de Blaise Pascal à Niels Bohr*)

by

G. Barton
Reader in Theoretical Physics, University of Sussex

Oxford New York Tokyo
OXFORD UNIVERSITY PRESS
1992

Oxford University Press, Walton Street, Oxford OX2 6DP
Oxford New York Toronto
Delhi Bombay Calcutta Madras Karachi
Petaling Jaya Singapore Hong Kong Tokyo
Nairobi Dar es Salaam Cape Town
Melbourne Auckland
and associated companies in
Berlin Ibadan

Oxford is a trade mark of Oxford University Press

Published in the United States
by Oxford University Press, New York

© Hachette, 1989, 79 boulevard Saint-Germain, 75006, Paris
Originally published in French by Hachette, Paris, 1989
This translation © G. Barton, 1992

A catalogue record for this book is available from the British Library

Library of Congress Cataloging in Publication Data
(Data available on request)
ISBN 0–19–853960–6
ISBN 0–19–8539770 (pbk)

Typeset by
Cotswold Typesetting Ltd., Gloucester
Printed in Great Britain by
Bookcraft Ltd., Midsomer Norton, Avon

Foreword
Alain Aspect
Laboratoire de Spectroscopie Hertzienne de
l'Ecole Normale Supérieure and Collège de France

The propagation of scientific knowledge beyond the circle of specialists is a difficult art. Who is best placed to exercise it has long been under debate. Some prefer a professional communicator even if totally untrained: only he is deemed sufficiently sensitive to the difficulties experienced by the general reader. Others maintain that only an expert in the field can avoid the snares that beset all popularization. I have had occasion to watch both kinds of writer at work in fields I know well, and can testify that both are capable of producing excellent texts, provided each knows his limitations. I have read very good articles by a media man with no previous knowledge of the field, but modest enough to have his work checked by experts (although, or perhaps because, he was a star performer on TV). I have also encountered eminent specialists with an equally virtuoso command of language directed to the layman.

Charles Ruhla has acquired the characteristic virtues of both kinds of writer. Here he addresses an audience already possessing some nontrivial acquaintance with basic science, but with little expertise in the subjects under discussion; and he is fully aware of the difficulties faced by the conscientious reader trained in classical physics, who wishes to tackle statistical physics or quantum physics.

The present book answers a very real need. For various reasons likely to operate for a long time yet, there are many people with a very creditable equipment of classical physics, but almost illiterate as regards the two pillars of modern physics. (It is not mere coincidence that students of the École Polytechnique are given a massive dose of statistical and of quantum mechanics as soon as they set foot in the School.) There are also many non-physicists with a taste for the sciences, who would like an introduction to physics at its most modern. To the curious of both kinds, Ruhla's book offers a most valuable tool for entry to a domain that is still much too far from being appropriately integrated into the culture of our society.

Preface

This book is dedicated to the memory of Mark Gusakow, professor at the Claude-Bernard University, and director of the Institute of Nuclear Physics, Lyons. His salty criticism, his comments full of insight, and his long-term encouragement have contributed much to bring this book into being.

Chance signifies the unpredictable, a long way apparently from science, whose essence is prediction. Yet even chance has its laws. By hard work over many years, physicists have explored it, substituting order of a subtle kind for the seeming disorder in the processes of nature.

The aim of this book is to present the laws of chance through a brief history running from Blaise Pascal and Pierre Fermat to Niels Bohr and Albert Einstein. We adopt the point of view of the physicist, always keen to introduce abstract concepts through concrete observations: on heads or tails, jammed telephone lines, billiards, expanding gases, the interference of light, and so on. In each case, basically qualitative arguments lead one to the probabilities characteristic of the phenomena. Those who prefer their reasoning quantified have not been forgotten: they will find the corresponding mathematical calculations in the appendices to the appropriate chapters.

The final chapters are devoted to an analysis of the recent experiments by Alain Aspect and his group on interference with a single photon, and on photon correlations. The interpretation of these experiments has implications for philosophy, and supplies some of the elements needed to answer a very basic question:

'Is chance merely an expression of our ignorance, or is it an inherent characteristic of natural phenomena?'

A better gift from physics to philosophy could hardly be imagined.

The approach I follow has been developed gradually, in a framework of Continuing Education for which I have been responsible at the University of Lyons since 1972. Faced with friendly and well-motivated audiences, I have had the opportunity to ring the changes on physics of many different kinds: practical, theoretical, didactic, and philosophic. Elements for a synthesis accumulated over the years, and it became very tempting to try and arrange them in some kind of order. But books are not written in isolation. First I had to find publishers. This was soon done: I am grateful to them for their welcome, and even more so for their patience. Next, I had to conform to the golden rule of the series in which this book originally appeared: no, or next to no mathematical detail, except perhaps in appendices; hence I had to look for a whole series of innovative analogies and demonstrations. Finally, I was set on leavening the mixture with some grains of fancy: readers making the arduous effort to understand the concepts of physics deserve a little entertainment from time to time.

I am of course wholly responsible both for the form and for the contents of the book, but I am also very mindful of all those who have contributed to it in different ways. My warmest thanks to Monsieur Aspect, for his foreword and his comments; Madame Cipan, for her careful reading of the MS; Mesdames Couchoud and Graveron, for many of the computer calculations; Messieurs Briguet, Caldero, Chanut, Gauthier, Jouanisson, and Rebouillat, for the demonstration experiments they arranged at my request; Messieurs Hernaus and Picard, who drew the diagrams.

My final thanks go to Monsieur de La Palice, whose proverbial logical faculties have helped me enormously in understanding and explaining quantum mechanics.

Lyons   C.R.
March, 1992

Note to the Reader

The values of physical and chemical constants used in this work are those recommended in 1987 by the International Union of Pure and Applied Physics.

Contents

Democritus. Greek philosopher; born around 460 BC in Abdera in Thrace, and said to have lived for more than a hundred years. He is the first to envisage atoms as small particles of matter, invisible and indivisible. Distributed randomly in space, the atoms are alike, because 'like attracts like'; they are the constituents of the familiar objects around us. These remarkable ideas about the structure of matter are associated with some rather rudimentary astronomy: the whole system is brought under suspicion by the astronomy of Aristotle, which comes later and can do much more. But the atomism of Democritus is revived very successfully by Dalton in 1806 (Chapter 4).
(Anderson-Viollet)

1

The children of Democritus
(prediction in science)

Everything that exists in the Universe is the fruit of
Chance and of Necessity
Democritus as quoted by Jacques Monod

Though our quote may not be word-perfect, it certainly expresses an idea to which Democritus was deeply committed, keenly aware as he was that both chance and necessity are essential characteristics of the phenomena of nature. However, succeeding physicists, from the Renaissance onwards, concentrated first and foremost on just one of the two, namely on necessity. The physics of chance started to be taken seriously only around 1650, with the theory of games; and only by slow progress and development did it achieve the primacy it now holds in contemporary physics. To appreciate the full import of this evolution, one must be prepared to start by confronting two rather philosophical questions: what is science for? and how does science function? We start with the first question.

1.1 Prediction = science

It is often thought that science is an explanation of the world. Though this is an important feature, it is not the most characteristic: *the overriding priority in science is prediction.*

As an example we consider eclipses, recalling that some primitive civilizations interpret them as actions of a malign divinity, enemy of the sun. This interpretation is certainly an explanation, but it generates no predictions; therefore it is not science. By contrast, the old Chaldeans had observed that eclipses occur regularly at an interval of 18 years, called the Saros cycle. The recognition of this cycle explains little beyond what is implied simply by the notion of regularity; on the other hand, it furnishes a very effective method of prediction. Therefore it is already part of science, albeit at its most elementary level, that of (mere) empiricism. Since then we have learnt to do rather better.

All methods of scientific prediction follow the same scheme. Its object is to construct a logical sequence leading from situation 1 to situation 2:

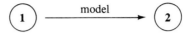

This sequence takes the form of a mathematical model, especially in the physical sciences, which are very thoroughly quantified. For example, the law of falling bodies, namely

$$z = \tfrac{1}{2}gt^2,$$

constitutes a model that allows one to follow the motion of a stone dropped from the top of a tower (situation 1), until it hits the ground (situation 2).

1.2 Model = scientific knowledge

Even though the primary function of science is to predict, it does also yield scientific knowledge as one of its particularly fascinating by-products. This then opens the door to philosophical debates about what the mathematical model actually represents.

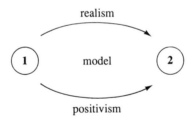

One school of thought, *realism*, to which Einstein belonged, postulates that the mathematical model is the image of something that really exists. The image itself is admittedly imperfect, but the fact that the model works guarantees the existence of some objective reality independent of the observer. Roughly speaking this implies that I believe in the existence of a meteor even if it does not land on my own head.

The *positivist* school of thought, to which Bohr belonged, holds that the only role of the mathematical model is to establish relations between observable entities. But it asserts that one can neither confirm nor deny the existence of any objective reality independent of the observer; indeed it claims that we cannot even tell whether this question has any meaning. Roughly speaking this implies that the stop-watch readings that time a runner are merely a method for relating his starting and his finishing times. They could be recorded perfectly well even if one never sees the runner, whose existence at this stage of the argument becomes somewhat mythical.

Fortunately, realists and positivists cooperate smoothly in the advancement of science, that is in making predictions; but they ascribe mutually contradictory meanings to the activity that achieves this.

1.3 Prediction in science

Heirs to Democritus and to his celebrated dictum, physicists have developed two different methods of prediction.

1. *Deterministic theories* describe systems where 'complete knowledge of the initial conditions and of the interactions allows one to predict the future exactly'. Celestial mechanics is perhaps the most beautiful example of a deterministic theory; convincing evidence for this view can be gathered by visiting a planetarium. The apparatus, heavily computerized, contains as input the observations of astronomers past; and as part of its program, the laws of Newtonian mechanics. Its predictive power is overwhelming. No sooner asked, the operator will show you the configuration of the heavens at noon on the first of January of the year 2000.

2. *Probabilistic theories* describe systems where 'the future cannot be predicted exactly'. For example, throwing a die I know that there are six possible outcomes, but I have no way of predicting which face will show uppermost. Therefore I assert that the end-result has come about by chance, which introduces the concept of *unpredictability* to describe dicing. We seem to be far removed from science, which is quintessentially predictive. But the creative imagination of mathematicians has surmounted this difficulty at least in part, by establishing 'laws of chance', that is laws for calculating probabilities. Three different kinds of prediction then become possible and useful.

- The first kind applies when the number of events is very large. If for instance I throw a die 6000 times, then I can predict that the number of 4's will be in the interval [913, 1087], at a confidence level of 0.997, which amounts to near certainty.

- The second kind of prediction applies to single events whose probability is either very high or very low. For example, if I travel by train for a whole year, the probability of avoiding an accident is 0.999 999, and the probability of a fatal accident is 0.000 001. Consequently I reckon the risk small enough to be acceptable, that is I take the train without much apprehension.

- The third kind of prediction relates to a single event with several possible outcomes, the probabilities for which are different. For example, if I throw two dice, the probability of a total of 7 is $\frac{1}{6}$, while the probability of a total of 2 or of 12 is only $\frac{1}{36}$. Hence 7 is a better bet than 2 or 12, which is by no means evident *a priori*.

To summarize, a deterministic theory allows one to predict the only evolution possible for a system, while a probabilistic theory allows one to predict the several different possible evolutions, and to assign a probability to each.

1.4 The hybrid case

Gunnery proves useful in helping us to understand the relation between deterministic and probabilistic theories (Fig. 1.1). If a gun is fired many times, always laid the same way, and always using the same type of shell with the same charge, one notices that the shells do not all land at the same point; rather the impacts are distributed at random over a small area, called the 'scatter zone'. Artillerymen are quite familiar with this phenomenon, and have tabulated it for every make of gun. In some artillery handbooks one even finds the following definition: 'Scatter is said to occur when the shell deviates from its trajectory and lands at a point different from its point of impact.' On recovering one's breath from such a picturesque definition, one realizes that it is easily corrected by writing 'average (or mean) trajectory' and 'mean (or barycentric) point of impact'.

The *deterministic features* of the process can be ascertained exactly. The mean point of impact *I* is perfectly calculable once we know the initial velocity v, the elevation α, and the density of air, ρ. The calculation, using the equations of classical mechanics, features two forces, gravitation and air-resistance.

Probabilistic features enter because in actuality v, α, and ρ cannot be determined with unlimited exactitude. There are small fluctuations from one discharge to the next, reflecting small variations at various levels:

- the mass and the machining of the shell;
- the powder charge;
- the elevation;
- air temperature and turbulence.

It is these small deviations from average conditions that are responsible for the scatter zone. The example shows that deterministic and probabilistic

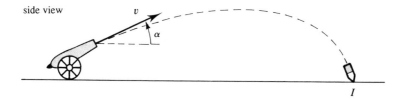

side view

v

α

I

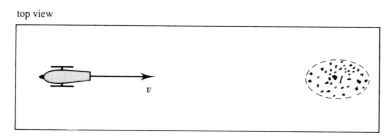

top view

v

Fig. 1.1 Basic ballistics.

theories are not mutually exclusive; on the contrary, they are in effect complementary. Situations differ only in the relative importance of the two. Determinism dominates in the planetarium, and probabilities in the fall of the dice. The problem of the gunner is a hybrid case displaying both aspects quite clearly; it is a good picture of the macroscopic world, being deterministic as regards the mean values of the measured quantities, and probabilistic as regards the scatter of the results.

One obvious question that comes to mind is whether the scatter can be reduced. The history of gunnery is nothing but the history of progress in the technology for precise control of the initial conditions, from the early bombards to the guns firing now. Every improvement lessens the deviation of the actual from the ideal fully deterministic trajectory, which in effect constitutes a limit. From this point of view, undiluted determinism plays the same role in physics generally as the absolute zero plays in the narrower domain of thermodynamics. We are persuaded that, objectively speaking, it exists, and believe that we shall get nearer and nearer to it, though without ever actually attaining it.

The discussion just presented is an accurate reflection of the views that govern classical physics, such as would have been acquired by a student towards the end of the nineteenth century. The objectives envisaged include the following.

1. On the one hand, there is the study of processes whose initial conditions

seem well-determined; such study would lead to deterministic theories like mechanics, electromagnetism, and classical thermodynamics.

2. On the other hand, there is the study of processes where one knows little or nothing about the initial conditions; such study would lead to probabilistic theories like the theory of games, the kinetic theory of gases, and statistical mechanics. Indeed one might well describe this domain as the realm of chance in its classical sense; we shall study it in Chapters 2 to 5.

This cosy understanding between the two approaches was to suffer a rude shock in the early twentieth century; as a result, probabilistic physics would evolve and chance extend its jurisdiction very considerably.

1.5 The limits of classical determinism

The essentials of the problem relate to the question whether a limit actually exists. There are two different reasons why the question must be reopened.

1. The behaviour of certain macroscopic systems is crucially affected by changes in the initial conditions that are quite minute. Such behaviour displays all the symptoms of randomness, though in fact it is only pseudo-random; the expression 'deterministic chaos' has been coined to describe it.

2. The behaviour of atomic systems shows that there are inescapable bounds to the precision with which initial conditions can be assigned. These are the bounds imposed by the Heisenberg inequalities, and they lead to a probabilistic theory, namely to quantum mechanics.

Both points constitute breathtaking developments in the physics of chance, and we shall study them in Chapters 6 and 7.

Finally, Chapter 8 will use the Einstein–Podolsky–Rosen paradox to discuss the great debate in present-day physics between chance and classical determinism, with our conclusions presented in Chapter 9.

Pierre Simon de Fermat. French mathematician; born in Beaumont-de-Lomagne, died in Castres (1601–1655). (Palais de la Découverte)

Blaise Pascal. French mathematician and philosopher; born in Clermont-Ferrand, died in Paris (1623–1662). (Palais de la Découverte)

Christiaan Huygens. Dutch astronomer, mathematician, and physicist; born and died at the Hague (1629–1695). (Photothèque Hachette)

Jakob Bernoulli. Swiss mathematician; born and died in Basle (1654–1705). (Boyer-Viollet)

These four men, all brilliant mathematicians, are the founding fathers of the theory of probability. Though there are famous names amongst their predecessors (Pacioli, Cardan, Galileo), it would be fair to claim that the calculus of probabilities stems from letters between Fermat and Pascal in 1654: Fermat contributes his results in combinatorics, and Pascal his celebrated *Treatise on the arithmetic triangle*. Huygens, informed about this correspondence, develops an interest in such problems, and in 1657 publishes his *De ratiociniis in ludo aleae*, the first treatise on the calculus of probabilities. Later, Jakob Bernoulli takes a hand by proving the law of large numbers, known nowadays as 'Bernoulli's theorem'; but it is published only in 1713, in his posthumous work entitled *Ars conjectandi*.

2

The laws of chance (the theory of probability)

Freedom is the right to do anything that the laws permit
Montesquieu

It seems at first that there is incompatibility between chance, which represents absolute randomness, and law, which is the symbol of enduring regularity. To the mathematical physicists of the second half of the seventeenth century (Fermat, Pascal, Huygens, Bernoulli) belongs the credit for realizing that even randomness could be confined within an appropriate framework; such was the birth of the theory of probability.

2.1 The experimental roots of the notion of probability

Philosophical interest was certainly not the only source of probability theory. One can well imagine that it was stimulated just as much by a more or less conscious search for some certain (but in fact illusory) method for success in gambling. Hence we should not be surprised that the first such theory to emerge was the theory of games, and that very symbol of randomness is the throw of the dice.

Experimental investigation with *just one die* is simple, and very instructive. It enables us at once to introduce some basic concepts, namely

- the *frequency* m, that is the number of throws producing a given result (say 4);
- the *relative frequency* m/n of this result (4), where n is the total number of throws.

It is very interesting to follow the evolution of the relative frequency m/n as a function of n; to this end we have made as many as 400 throws. The results are shown on the graph, with n plotted horizontally in steps of 12, and with m/n plotted vertically. The relative frequency m/n is a discontinuous variable, but the sequence of the points conveys quite a clear idea of the evolution we are looking for (Fig. 2.1).

It is evident from the graph that the observed relative frequency fluctuates, and that the amplitude of these fluctuations falls as the number n of throws rises. The ratio m/n approaches $\frac{1}{6}$, and we are entitled to think that this limit would be reached if n became infinite.

On considering a single throw, *intuition* leads us to believe that none of the six faces of the die is privileged *a priori*. Even though it is impossible to predict the result of just one throw, it seems reasonable to think that, in many throws, the number 4 will 'on average' occur once in six throws, and that its frequency m/n will tend to $\frac{1}{6}$.

Our definition of *the probability P* stems from this accord between experience and intuition. We shall write it as a ratio:

$$P = \frac{\text{number of possible outcomes where the desired event occurs}}{\text{total number of possible outcomes}}.$$

For a single throw, the probability P that the number 4 appears is $\frac{1}{6}$.

This formulation contains three hidden assumptions, which are worth explicating.

1. As our quantitative measure of chance, we have chosen a magnitude P without physical dimensions, having a value between 0 and 1, and capable of attaining these extremes. The extreme 0 corresponds to cases that are impossible (throwing 7, 8, 9, ...); the extreme 1 corresponds to certainty (throwing 1 or 2 or 3 or 4 or 5 or 6).

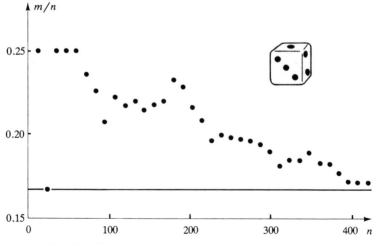

Fig. 2.1 The measured relative frequency tends to a limit.

2. The special case of 'equally probable' phenomena serves to define probability! Are we begging the question? Not really. As always in physics, it is intuition that leads one to the concepts. The concept of *a priori* probability is suggested by the apparent equivalence between the six faces of the die. Of course it is quite possible that after very many throws the relative frequency m/n of producing 4 tends to a limit different from $\frac{1}{6}$. If this does seem to be happening, it is prudent not to jump to conclusions too soon. For instance, in the example quoted in Chapter 1, where we threw the die 6000 times, one would need to observe a deviation of at least 9% before believing that the effect is really significant.† If the deviation is indeed significant, one concludes *a posteriori* that the die is biased. This comment stresses the subtle distinction between *a priori* probability, which relates to a die unbiased by assumption, and *a posteriori* probability, which characterizes the actual die. The throw of a biased die is still a random event, but it is to the player's advantage to know that the *a posteriori* probability exceeds $\frac{1}{6}$; by contrast, the uninformed player is aware only of the *a priori* probability, which is $\frac{1}{6}$.

3. It is necessary to assume that two successive throws are totally independent, or in other words that *the die has no memory*. In practice this condition is satisfied only if the die is given enough momentum to roll on the table. In that case the probability of throwing a 4 is indeed $\frac{1}{6}$ every time, even if it so happens that 4 has not shown up once in the last thirty throws. This assumption, though it stems from the rational analysis of games of chance, runs counter to primitive intuition. Who has not heard a lottery player declare that 4 has not come up in the last two months, wherefore there is an 'excellent chance' of its coming up this week? The idea though illusory is very common. In fact, if 4 has an 'excellent chance' of coming up, then this can mean only that its probability is not small. At every draw seven balls are picked from a total of 49; calculation then yields a probability of 0.152 for the appearance of any given number, and this probability is not affected by earlier draws. As to the probability of winning first prize, it is very low indeed, because it requires the appearance of six numbers chosen in advance, and calculation shows that it amounts to only 0.000 000 072. But dreams cannot be costed!

2.2 Random and pseudo-random phenomena

The preceding section showed that throwing dice is indeed unpredictable, whence it is a true random process provided one proceeds with care. On the

---

† To derive this estimate, one first calculates the mean-square deviation from the formula $\sigma = (nab)^{1/2}$ (Section 2.4), and then applies the geometric criterion $\pm 3\sigma$ stemming from the Gaussian distribution (Section 2.6). The estimate is an approximation that matches the shape of a binomial distribution for $n = 6000$ to the shape of a Gaussian distribution. It then turns out that here the interval corresponds to 8.7% at a confidence level of 0.997.

other hand the process is slow, and makes the experimental verification of the laws of chance time-consuming and tedious. One is tempted to look for a faster method, and the computer would appear to be an ideal tool because of its speed. Is this altogether true?

By using an appropriate rule one can program a sequence of numbers a_n such that $0 \leq a \leq 1$, defined in terms of a starting value a_0 which is chosen arbitrarily (see Appendix 1). The sequence of the a_n is deterministic, since it can be reproduced as often as one likes from the same starting value a_0. Every one of the elements a_n is fully determined by the preceding element. Yet the resulting sequence displays every sign of randomness: an operator ignorant of the recurrence rule and of a_0 has practically no chance of discovering the deterministic key to the process. In such cases one speaks of *pseudo-random* series, and the sequence of the a_n allows one to simulate the throw of a die to excellent effect; moreover it has the advantage of great speed (see Appendix 1).

Here, as an example, is the squence of the first 30 throws, constructed from $a_0 = 0.5$:

$$5, \quad 1, \quad 4, \quad 6, \quad 1, \quad 5, \quad 1, \quad 1, \quad 4, \quad 6, \quad 4, \quad 6, \quad 3, \quad 6, \quad 5,$$
$$5, \quad 5, \quad 3, \quad 3, \quad 5, \quad 4, \quad 2, \quad 6, \quad 3, \quad 1, \quad 3, \quad 4, \quad 5, \quad 2, \quad 3.$$

Look next, as in the preceding section, at the evolution of the relative frequency m/n of throwing a 4 (Fig. 2.2). The simulation was implemented on a microcomputer, collecting the throws into groups of 12 in order to make the variation of m/n easier to follow.

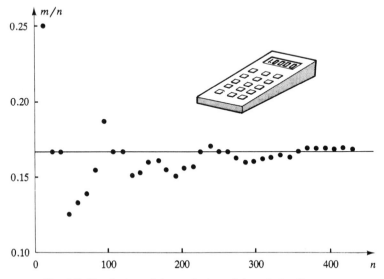

Fig. 2.2 Simulation of the evolution of the relative frequency.

Comparison between the evolutions of the observed relative frequencies in Figs 2.1 and 2.2 is convincing evidence for the excellence of the simulation; the limit $\frac{1}{6}$ is the same for the random process (Fig. 2.1) as for the pseudo-random process (Fig. 2.2).

2.3 Total and joint probabilities

We introduce two important theorems without proof, by appeal to two specific examples.

Problem 1. Draw one card at random from a pack of 52.

- What is the probability of drawing a diamond?
  Answer: $13/52 = \frac{1}{4}$.

- What is the probability of drawing a heart?
  Answer: $13/52 = \frac{1}{4}$.

- What is the probability of drawing a red suit?
  Answer: $26/52 = \frac{1}{2} = \frac{1}{4} + \frac{1}{4}$.

The theorem of total probabilities: If an event C occurs when either an event A or an event B occurs, where A and B are mutually independent, then the probability that C occurs is the sum of the two separate probabilities for A to occur and for B to occur.

Accordingly we write

$$P(C) = P(A) + P(B).$$

One particularly interesting case is that where B is the event 'not A' (written $\bar{A}$), that is the event contrary to A. For instance, let A be the occurrence of a diamond in a single draw, whence $P(A) = \frac{1}{4}$. Then $\bar{A}$ is the drawing of a heart or of a spade or of a club, and $P(\bar{A}) = \frac{3}{4}$. The event C is the drawing of any card, and we have

$$P(C) = P(A) = P(\bar{A}).$$

But it is certain that *some* card will be drawn, so that $P(C) = 1$, whence

$$P(A) + P(\bar{A}) = 1.$$

This relation entails that all probability distributions are normed to 1.

Problem 2: Two cards are drawn simultaneously, one from each of two different packs.

- What is the probability of drawing a red card from the first pack?

Answer: $26/52 = \frac{1}{2}$.

- What is the probability of drawing a red card from the second pack?
  Answer: $26/52 = \frac{1}{2}$.
- What is the probability of drawing two red cards in the joint operation?
  Answer: $\frac{1}{2} \times \frac{1}{2} = \frac{1}{4}$.

The theorem of joint probabilities: If an event C results from the simultaneous occurrence of an event A and of another event B, where A and B are mutually independent, then the probability that C occurs is the product of the probabilities of occurrence of A and of B separately.

The assumption that A and B are mutually independent is basic to this theorem. The result is written

$$P(C) = P(A) \times P(B).$$

It is on account of this theorem that the President and the Prime Minister of France travel in two different aeroplanes even when they are going to the same place: the probability that both politicians will disappear simultaneously and by accident is then minute (less than 10^{-12}).

2.4 The binomial distribution

The game of heads or tails serves to introduce the notion of a *discontinuous statistical distribution*, with the *binomial distribution* as a simple example. The calculation of the *a priori* probabilities is elementary, because there are only two possibilities, heads (written A) and tails (written B); both have probability $\frac{1}{2}$. The possible outcomes of the first three trials are enumerated as follows:

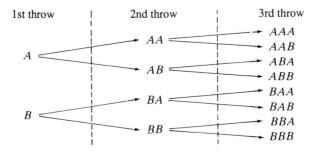

1st throw	2nd throw	3rd throw
		AAA
	AA	AAB
A		ABA
	AB	ABB
		BAA
	BA	BAB
B		BBA
	BB	BBB

The statistical distribution is a convenient representation of these results. To exploit it we shall need the following notation:

n, an integer, is the number of tosses;

m, an integer, is the number of outcomes A (heads) in n tosses (it is a discontinuous random variable);

a is the probability of the outcome A (heads), i.e. $\frac{1}{2}$;

b is the probability of the outcome B (tails), i.e. $\frac{1}{2}$.

We notice straightaway that $a+b=1$, since every toss of the coin produces either heads or tails.

The following table shows the results of three successive tosses:

n	Total number of possible cases	Number of favourable cases		Probability $P(m)$ that A occurs m times in n throws	Sum of all the probabilities for given n (norming condition)
1	2	$m=0$	1	0.5	1
		$m=1$	1	0.5	
2	4	$m=0$	1	0.25	1
		$m=1$	2	0.5	
		$m=2$	1	0.25	
3	8	$m=0$	1	0.125	1
		$m=1$	3	0.375	
		$m=2$	3	0.375	
		$m=3$	1	0.125	

For a given value of n, the statistical distribution for the game of heads or tails consists in the set of all pairings $(m, P(m))$. The distribution is discontinuous, and it can be represented by a graph which is just a sequence of discrete points (Fig. 2.3). We have refrained deliberately from joining these points with broken lines, which might guide the eye but have no physical significance.

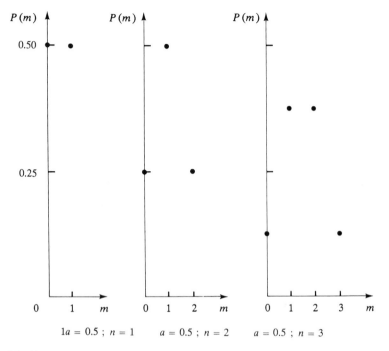

Fig. 2.3 The probability of heads in one, two, and three throws of one and the same coin.

Recourse to mathematical analysis seems to become indispensable as soon as n becomes large. In Fig. 2.4 we have, after an initial change of scale, drawn the distributions corresponding to $n = 5$, $n = 10$, and $n = 15$. Notice that the highest probabilities occur near $\frac{1}{2}n$, and the lowest near 0 and n.

Finally, after a further change of scale, we have drawn the distribution for $n = 30$ (Fig. 2.5). The rescaling makes it easy to see the limit that corresponds to infinite n: it is a continuous distribution in the form of a symmetric bell curve, which we shall soon be calling the normal (or Gaussian) distribution (Section 2.6).

Even short of infinite n, we shall see that for very large n the binomial distribution can furnish an excellent account of the irreversible (Gay-Lussac) expansion of gases (Section 5.1).

The symmetric binomial distribution is merely one special case, corresponding to $a = b = 0.5$. But in fact the only general condition on a and b is $a + b = 1$, and we are quite free to investigate other cases. With dice for instance, the probability of throwing a 4 is $a = \frac{1}{6}$, and the probability of throwing not-4 is $b = \frac{5}{6}$. In Fig. 2.6 we have drawn the binomial distribution for eight successive throws of the same die; the discontinuous random variable

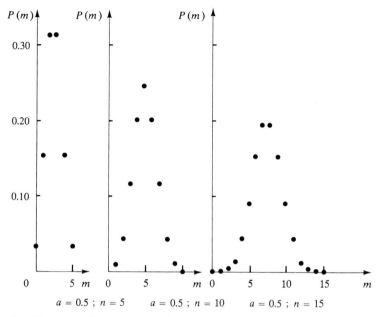

Fig. 2.4 The probability of heads in five, ten, and fifteen throws of one and the same coin.

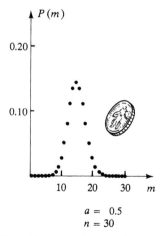

Fig. 2.5 The probability of heads in thirty throws of one and the same coin.

m counts the number of 4's. This distribution is highly asymmetric; and it shows that one must not confuse the most probable value of m (1 in this case) with the average $\bar{m} = na$ (1.33 in this case: see Appendix 2). In fact we are already in a position to anticipate that, when n is large, these asymmetric distributions can be represented through a very simple approximation, namely through the Poisson distribution (Section 2.5).

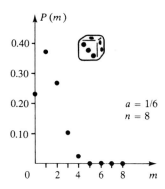

Fig. 2.6 The probability of throwing a 4 in eight throws of one and the same die.

The binomial distribution has a very simple mathematical form, and since its application is wholly unrestricted it covers all the several cases which we have cited. It reads

$$P(m) = C_n^m a^m b^{n-m}.$$

Here, n is the total number of trials; m, a discrete random variable, is the number of times event A occurs; a is the probability that event A occurs; b is the probability that the event contrary to A occurs, whence $a + b = 1$; C_n^m is the number of combinations one can form from n objects taken m at a time, and is given by

$$C_n^m = \frac{n!}{m!(n-m)!}, \quad \text{with } m \leq n;$$

finally $P(m)$ is the probability that the event A occurs m times (in n trials).

Notice that this distribution is fully specified by two parameters, n and a; but generally one prefers to define it through two other and rather more informative parameters. These are

- the *mean value* or *average*: $\bar{m} = na$, also called the *expectation value*;
- the so-called *standard deviation*: $\sigma = (nab)^{1/2}$.

All these results are established in Appendix 2.

2.5 The Poisson distribution

The approximation '*a* small, *m* middling, and *n* not too large' is useful as a mnemonic. But the truly operative condition relates to the mean value $\bar{m} = na$, which must not exceed a few units (i.e. it must not be of order of magnitude greater than 1). Typically one meets this condition with values $a \leq 0.01$, $0 \leq m \leq 10$, and $n \geq 100$. Then it becomes possible to find a simple approximation to the binomial distribution: this is the *Poisson distribution*, which reads

$$P(m) = \frac{\bar{m}^m}{m!} \exp(-\bar{m}),$$

where m, the number of times the event A occurs, is an integer-valued, and therefore discontinuous, random variable, and $\bar{m}$ is the mean (average) value of the random variable m. Though $\bar{m}$ enters via the relation $\bar{m} = na$, one may disregard this fact; in view of the expression for $P(m)$, the mean value $\bar{m}$ then becomes the only parameter in the distribution. The value of $\bar{m}$ must be of order 1, but it can be either integer or non-integer. Finally, $P(m)$ is the probability that the event A will occur m times in the course of the experiment.

The sequence of pairs $(m, P(m))$ amounts to a discontinuous distribution, namely to the Poisson distribution appropriate to a mean value $\bar{m}$. All the properties of the Poisson distribution are established in Appendix 3.

Telephone jamming furnishes a concrete example of the distribution through the following problem. In eight working hours a commercial telephone line receives, on average, n calls lasting two minutes each. Study the performance of this line for the three cases $n = 24$, $n = 240$, and $n = 1200$.

The choice of two minutes for every call is not too unrealistic a model, allowing for callers of all kinds: chatty, concise, impatient, calm, etc.; making the model more elaborate would not change the solution in any basic way. The number of calls attempted in a two-minute interval (event A) is represented by the random variable m, and the key to the problem is the average $\bar{m}$ of m. We have the general relation

$$\bar{m} = \frac{n}{8 \times 30},$$

whence

$$n = 24 \quad \text{gives } \bar{m} = 0.1,$$
$$n = 240 \quad \text{gives } \bar{m} = 1,$$
$$n = 1200 \quad \text{gives } \bar{m} = 5.$$

These values are indeed no greater than of order 1, which justifies appeal to the Poisson distribution. The corresponding distributions read

$$P(m) = \frac{(0.1)^m \exp(-0.1)}{m!} \quad \text{for } \bar{m} = 0.1,$$

$$P(m) = \frac{\exp(-1)}{m!} \quad \text{for } \bar{m} = 1,$$

$$P(m) = \frac{5^m \exp(-5)}{m!} \quad \text{for } \bar{m} = 5.$$

They are drawn in Fig. 2.7.

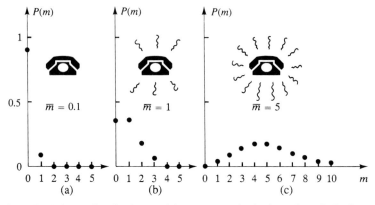

Fig. 2.7 The Poisson distribution enables one to study the jamming of telephone lines.

The only scenario satisfactory to a caller is that there should be no other caller on the line in the interval under consideration (two minutes); otherwise his own call will be rejected. This scenario corresponds to $P(0)$, and we examine three possibilities.

- $m = 0.1$: this yields $P(0) = 0.9$. In other words there is a chance of 90 in a 100 of finding the line free after waiting no more than two minutes. This scenario is acceptable (Fig. 2.7a).
- $m = 1$: this yields $P(0) = 0.37$. In other words your chances of finding the line free are barely above one in three. This scenario is acutely irritating (Fig. 2.7b).
- $m = 5$: this yields $P(0) = 0.007$, a very low value. The chances of getting through are minuscule. From the graph (Fig. 2.7c) we see that the most probable cases are four or five calls in two minutes. They all get in each other's way, and everyone is cross. This is the quintessential swamped line,

such as can occur when legions of television viewers wish to register enthusiasm or disapproval.

We have got well beyond simple game theory. The economic implications of the Poisson distribution are evident: it is this distribution that allows us to foresee how many telephone lines will be needed to meet heavy demand. Through this special case we have just uncovered a problem very widespread in industry, called the *queuing problem*. As to its physical characteristics, some further comments are in order.

1. The situation $m = 1$ corresponds to the maximum theoretical capacity of the line, on the assumption that the calls arrive *periodically* at two-minute intervals. One would then have 240 calls lasting two minutes each, accounting for 480 minutes, that is for an eight-hour working day. But in fact the calls arrive *at random*, entailing an efficiency of only 37% and a loss rate of 63%. Indeed $\bar{m}$ must drop to 0.1 before one reaches acceptable working conditions, that is an efficiency of 90% and a loss rate of 10%; in other words the call rate must drop by a factor of 10. Hence the following rule of thumb:

The efficiency of a detector drops by a factor of 10 when the signals change from periodic to random.

The rule applies not only to telephone calls but also to particle counters.

2. The Poisson distribution depends on just one parameter, namely on $\bar{m}$; this applies both to the calculation of the probabilities through

$$P(m) = \frac{(\bar{m})^m \exp(-\bar{m})}{m!}$$

and to the calculation of the standard deviation

$$\sigma = (\bar{m})^{1/2}.$$

In Fig. 2.7c, the graph for $m = 5$ is an almost symmetric bell curve. This shows in passing that the Poisson distribution approaches the normal (Gaussian) distribution (see Section 2.6 below) when $\bar{m}$, and therefore n, become very large (even if a is very small).

To end this section on an entertaining note, let us envisage a cake containing 100 raisins and cut into 20 equally thick slices. The average number of raisins per slice is $m = 5$, but the probability of receiving a slice containing precisely five is only 0.17. There are other possibilities: the slice with no raisins at all for instance (probability 0.0067), balanced fortunately by the slice with ten raisins (probability 0.018). Such are the lessons taught by the Poisson distribution for $\bar{m} = 5$.

2.6 The Gaussian (or normal) distribution

This is a *continuous* statistical distribution, called 'normal' because it is very often encountered in nature. It is so important that the name of its inventor has been immortalized in the word 'Gaussian' (which can be either adjective or noun). We propose to introduce it in concrete fashion by describing an experiment easily done in any modern laboratory.

Noise in a vacuum tube consists of a succession of electric pulses random both in amplitude and in frequency; it arises from the numerous fluctuations of the electron beam traversing the tube. By feeding it into a suitable amplifier one obtains a noise generator, and we aim to study the amplitude distribution of the output signal. Such a distribution is called a *noise spectrum*; one uses the experimental layout sketched in Fig. 2.8.

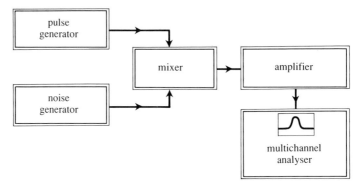

Fig. 2.8 The experimental study of noise in a vacuum tube.

We start by identifying the role of all the elements of this layout.

- The pulse generator delivers periodic signals of constant amplitude (Fig. 2.9a).
- The noise generator delivers a continuously varying voltage, random both in amplitude and in frequency (Fig. 2.9b).
- The mixer accepts the signals from both generators, and delivers as output a sequence of pulses at equal time intervals, but whose amplitude is random (Fig. 2.9c).
- The pulses from the mixer are amplified and then fed into a multichannel analyser. The role of the multichannel analyser is to classify the pulses it receives, and to count them.

- The pulses are classified according to their voltage. The amplification is chosen so that the full gamut of 100 channels spans a potential difference of 12 V, plotted horizontally. Hence each channel represents a slice of 0.12 V.
- The counts are plotted vertically. These are the numbers of events (pulses) in each slice of width 0.12 V, that is in each channel.

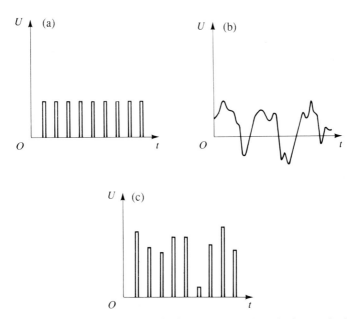

Fig. 2.9 The signals used to display noise in a vacuum tube (*t* is time; *U* is the signal amplitude).

The pulse generator supplies well-controlled pulses, at the rate of 50 a second; after amplification all these pulses rise to 6 V, with fluctuations less than 0.06 V. Hence they will all register in channel 50, the amplification having been adjusted carefully to achieve just this. In the language of the physicist, one would say that 'the spectrum of the generator is a single peak straddled by just one channel'.

To the regular pulses from the pulse generator we now add the random signals from the noise generator. After amplification this yields a very broad distribution, centred on channel 50 but stretching over more than 30 channels. The role of the pulse generator is 'to shift the zero level' in order to match the noise signals, which can be positive or negative, to the input requirements of the multichannel analyser, which accepts only positive input. Without such a stratagem we would obtain a distribution peaking in channel 0 rather than channel 50, and we would lose all the noise signals that are negative.

The experimental results are shown in Fig. 2.10. On the horizontal axis we plot the channel number, which is proportional to the signal strength (0.12 V per channel); vertically, we plot the number $N(x)$ of the events in each channel, proportional in turn to the probability density $\rho(x)$ (see below). Though the noise voltage varies continuously, the results appear as a sequence of discrete points, because the multichannel analyser groups the values into slices 0.12 V wide. Nevertheless there are enough points to suggest the general profile of the distribution, which is that of a symmetric bell curve.

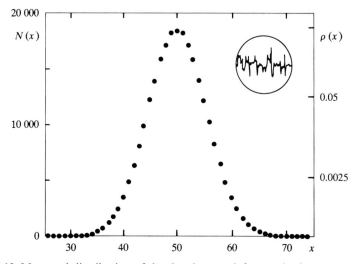

Fig. 2.10 Measured distribution of the signal strength from noise in a vacuum tube.

Mathematical smoothing techniques then yield the continuous distribution underlying our sequence of points. It is a function of the form

$$\rho(x) = \frac{1}{5.4(2\pi)^{1/2}} \exp\left(-\frac{(x-50)^2}{2 \times 5.4^2}\right).$$

The graph representing $\rho(x)$ agrees closely with the experimental values; we have here a good example of a very common continuous distribution, called a Gaussian. It reads

$$\rho(x) = \frac{1}{\sigma(2\pi)^{1/2}} \exp\left(-\frac{(x-\bar{x})^2}{2\sigma^2}\right),$$

where x is a continuous random variable, $\rho(x)$ is the probability density, $\bar{x}$ is the mean value, and σ is the standard deviation.

The use of a *probability density* depends on the fact that the variable x is continuous. The quantity $\rho(x)\Delta x$ is the probability of observing an event in the interval of width Δx between x and $x+\Delta x$. Finally we must stress that the mathematical form of $\rho(x)$ is by no means arbitrary: it arises on taking the limit of the binomial distribution when the number n of trials becomes very large (see Appendix 5).

The two parameters $\bar{x}$ and σ define the distribution completely, by prescribing the values of the two characteristics of the physical quantity we have chosen as our representative random process. In the experiment just described, this quantity is a random voltage fed into the multichannel analyser. Its mean value is 6 V (corresponding to $\bar{x}=$ channel 50), and its fluctuations are well described by the standard deviation, whose value is 0.65 V (corresponding to $\sigma=5.4$ channels). We emphasize the fact that in this example the observed fluctuations are by no means due to the measuring apparatus; rather they correspond in all essentials to the natural width of the phenomenon under study. This is easily proved by disconnecting the noise generator. Events then accumulate only in channel 50, which shows that the multichannel analyser warrants the voltage within a range of ± 0.06 V, ten times less than the standard deviation of the phenomenon we are investigating. Accordingly we have measured, to an accuracy of two significant figures, two characteristics of a random voltage having a large natural width: namely its mean value, 6.0 V, and its standard deviation, 0.65 V.

When we tackle the theory of the measurement of continuous variables in Chapter 3, we shall be in the opposite situation, needing to analyse phenomena whose natural width is narrow, by means of measuring apparatus introducing fluctuations that are much wider. Of course there are also hybrid cases where the two widths are comparable; this is a problem for metrological laboratories devising standards, but we shall not pursue it here.

The Gaussian distribution is *normed*. Mathematically speaking this means that

$$\int_{\infty}^{\infty} \rho(x)\,\mathrm{d}x = 1.$$

Physically speaking this relation gives the probability for finding the variable x in the open interval $(-\infty, +\infty)$. This probability is equal to 1, and the outcome is certain. In practice, of course, one is not dealing with the full open interval $(-\infty, +\infty)$; in our example of electronic noise, these limits would correspond to very rare signals of large if not infinite amplitude. Instead, one must adopt 'reasonable' limits on the amplitude, corresponding to similarly 'reasonable' probabilities. In order to define such a range of amplitudes, we use the standard deviation as a criterion, writing the limits as $\bar{x}-t\sigma$ and $\bar{x}+t\sigma$, where t is some positive number of order 1, called the safety factor. The probability of finding the variable x in this range is

$$W(t\sigma) = \frac{1}{\sigma(2\pi)^{1/2}} \int_{\bar{x}-t\sigma}^{\bar{x}+t\sigma} \exp\left(-\frac{(x-\bar{x})^2}{2\sigma^2}\right) dx.$$

It is the area under the Gaussian in the range $[\bar{x}-t\sigma, \bar{x}+t\sigma]$.

The geometric properties of the Gaussian distribution are shown in Fig. 2.11, together with a table giving values corresponding to some ranges. The most widely used range is, of course, $\pm\sigma$, corresponding to $t=1$ and to a probability practically equal to $\frac{2}{3}$. However, depending on the object in view, one may be led to choose other ranges. For example, $t=0.6745$ defines the *median range* used by gunners, and corresponds to one chance in two that the shell will land in this range. Similarly, $t=1.177$ corresponds to the half-width at half-height, and can serve to define *resolving power*. Finally, $t=3$ goes to the heart of the matter, because the probability of meeting an event outside the range $\pm 3\sigma$ is 0.003, which is very small.

In taking decisions that turn on random phenomena, the habits of particular disciplines and of particular countries are now beginning to be superseded by internationally agreed standards:

- $t=1.96$ (probability 0.95) in quality control;
- $t=2.58$ (probability 0.99) in the sciences;
- $t=3.29$ (probability 0.999) when some very large investment is at stake.

The Gaussian distribution has a quasi-universal character, being widely applicable not only in the physical but also in the natural and in the social sciences.

The best way to end this chapter is to set the reader the following modest exercise for the holidays.

Imagine a beach that can be reached by only one path. How are the holidaymakers distributed as regards their distance from the path?

Detailed analysis of the problem suggests that there exist

1. a basic governing factor, the end of the path, which it is usual to choose as the zero of the x-coordinate;

2. a whole host of secondary factors, each relatively unimportant on its own, but which, taken jointly, do have an appreciable influence on the behaviour of the visitors, leading them to choose some particular x-value for their position on the beach. A partial list might read thus:

   - physical factors affecting the ease of locomotion, like the position of the

t	$W(t\sigma)$	h_t/h_0
0.674	0.500=1/2	0.797
1	0.683	0.607
$1.177=\sqrt{2\ln 2}$	0.761	0.500=1/2
$1.414=\sqrt{2}$	0.843	0.368=1/e
2	0.954	0.135
3	0.997	0.011

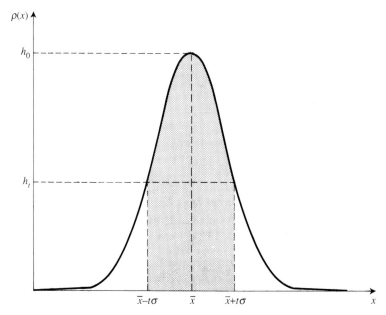

Fig. 2.11 Geometric properties of the Gaussian distribution. $W(t\sigma)$ is the probability of finding x in the range $[\bar{x}-t\sigma, \bar{x}+t\sigma]$.

sun in the sky, the smoothness of the sand, the weight of equipment carried, etc.;

- psychological factors like degrees of gregariousness or of misanthropy, greed (in view of an ice-cream vendor stationed at the path-end which one must pass anyway), the vigilance of parents with several children, and so on.

In general, it is difficult to define and to measure the influence of these factors taken singly; on the other hand, their overall result is quite easy to measure, and can hardly be overlooked by any seasoned observer: the distribution of the bathers is Gaussian, with a standard deviation rarely greater than a few hundred metres. This deviation represents the average effort that the average holidaymaker is prepared to expend in order to secure his personal square metre of hot sand.

Our example has not been chosen arbitrarily. Under the behaviour of the average holidaymaker there lurks an absolutely basic theorem, the *central limit theorem*, which will supply the framework of Chapter 3.

Appendices to Chapter 2

1 Computer simulation of the throw of a die

Microcomputers and scientific pocket calculators usually have a function RND (for 'random'), which supplies a pseudo-random sequence of numbers between 0 and 1. If by misfortune no such function is incorporated, one can very easily programme one for oneself. Here for example is the generating function suggested in the user's manual for the HP-34C calculator:

$$a_n = \text{FRAC}\,(9821a_{n-1} + 0.211\,327),$$

where FRAC stands for the fractional part.

This is a recurrence relation which constructs a sequence of numbers a_n, starting from an initial value a_0 called the 'germ', which the operator chooses at whim from the closed interval $[0, 1]$. The relation satisfies the *spectrum test*, which is a mathematical criterion guaranteeing that the numbers a_n do not bunch in any particular region of the interval $[0, 1]$. From the value of a_n one then determines the result of throw number n, call it j_n, through

$$j_n = \text{INT}\,(6a_{n-1} + 1),$$

where INT stands for the integer part.

As an illustration, here are the results from an HP-34C calculator with the choice $a_0 = 0.5$:

n	1	2	3	4	5	6	7	8	9	10	
a_n	0.711327	0.153794	0.622201	0.847348	0.016035	0.691062	0.131229	0.011336	0.542183	0.990570	...
j_n	5	1	4	6	1	5	1	1	4	6	...

The sequence can be extended up to $n = 1\,000\,000$ without any periodicity apparent in the results. Accordingly it simulates the throwing of a die very well.

2 The binomial distribution

Let a be the probability of event A, and b the probability of the event contrary to A. These definitions entail $a + b = 1$. We require the probability $P(m)$ that in a sequence of n trials A happens m times and B happens $n - m$ times, subject of course to the condition $m \leq n$.

One obvious possibility is that m events A happen first, followed by $n - m$ events B. The probability of this is $a^m b^{n-m}$. We have merely applied the theorem of joint probabilities (the events being supposed independent).

However, since the order of the events is not prescribed, all the possibilities for m appearances of A in the n trials are equally acceptable. The number of these possibilities is

$$C_n^m = \frac{n!}{m!(n-m)!},$$

a classic result in combinatorics.

Each of these possibilities has the same probability $a^m b^{n-m}$. It remains only to apply the theorem of total probabilities, which yields the requisite distribution as

$$P(m) = \frac{n!}{m!(n-m)!} a^m b^{n-m}.$$

This is called the binomial distribution because the numbers $P(m)$ are the successive terms in the expansion of the binomial $(a + b)^n$.

To calculate the two parameters $\bar{m}$ (the mean) and σ (the standard deviation), we shall use the 'moment-generating function'

$$\phi(t) = (at + b)^n.$$

Expanding the binomial $(at + b)^n$, one finds

$$\phi(t) = \sum_{m=0}^{n} C_n^m a^m b^{n-m} t^m.$$

In the term of order m we recognize the coefficient of t^m as the probability $P(m)$, whence

$$\phi(t) = \sum_{m=0}^{n} P(m) t^m.$$

- The first derivative of $\phi(t)$ leads us to $\bar{m}$, as follows. Express $d\phi/dt$ in two different ways:

$$\frac{d\phi}{dt} = \sum_{m=1}^{n} mP(m)t^{m-1}, \qquad \frac{d\phi}{dt} = \frac{d}{dt}[(at+b)^n] = na(at+b)^{n-1}.$$

Next, set t equal to 1:

$$\left[\frac{d\phi}{dt}\right]_1 = \sum_{m=1}^{n} mP(m), \qquad \left[\frac{d\phi}{dt}\right]_1 = na \quad \text{for } a+b=1.$$

The first equality is nothing but the definition of the mean value $\bar{m}$ of the discontinuous random variable m. The second equality then supplies the value of $\bar{m}$. Accordingly, $\bar{m} = na$.

- The second derivative leads us to σ. In order to find a parameter characterizing the width of the distribution, one cannot rely on the algebraic mean of the deviations $m - \bar{m}$, because the value of this mean is zero. Thus one is led to exploit the *variance*, namely the mean value of the squares of the deviations, which never vanishes. The *standard deviation* σ is defined as the square root of the variance; in other words

$$\sigma = \left(\sum_{m=0}^{n} P(m)(m-\bar{m})^2 \right)^{1/2},$$

or equivalently

$$\sigma = [\overline{(m-\bar{m})^2}]^{1/2}.$$

Since we know that the average of a sum is the sum of the averages, we have

$$\sigma = [\overline{(m^2)} - \overline{(2m\bar{m})} + (\bar{m})^2]^{1/2} = [\overline{(m^2)} - 2(\bar{m})^2 + (\bar{m})^2]^{1/2},$$

whence finally

$$\sigma = [\overline{(m^2)} - (\bar{m})^2]^{1/2}.$$

Before we can determine σ, we need $\overline{(m^2)}$, which will be supplied by $\phi(t)$. To this end, we express $d^2\phi/dt^2$ in two different ways:

$$\frac{d^2\phi}{dt^2} = \sum_{m=2}^{n} m(m-1)P(m)t^{m-2} = \sum_{m=2}^{n} m^2 P(m)t^{m-2} - \sum_{m=2}^{n} mP(m)t^{m-2},$$

$$\frac{d^2\phi}{dt^2} = \frac{d^2}{dt^2}[(at+b)^n] = \frac{d}{dt}[na(at+b)^{n-1}] = n(n-1)a^2(at+b)^{n-2}.$$

Again we set t equal to 1, and recall that $a+b=1$; hence

$$\left[\frac{d^2\phi}{dt^2}\right]_1 = \sum_{m=0}^{n} m^2 P(m) - \sum_{m=0}^{n} m P(m), \qquad \left[\frac{d^2\phi}{dt^2}\right]_1 = n(n-1)a^2.$$

The first expression equals $\overline{(m^2)} - \bar{m}$, and its value is given by the second. Accordingly,

$$\overline{(m^2)} - \bar{m} = n(n-1)a^2;$$

since $\bar{m} = na$, this entails

$$\overline{(m^2)} = n(n-1)a^2 + na, \qquad (\bar{m})^2 = n^2 a^2.$$

It follows that

$$\sigma = [n(n-1)a^2 + na - n^2 a^2]^{1/2} = [na(1-a)]^{1/2}.$$

Since $1 - a = b$, the end-result is very simple:

$$\sigma = (nab)^{1/2}.$$

3 The Poisson distribution

Starting from the binomial distribution

$$P(m) = \frac{n!}{m!(n-m)!} a^m b^{n-m},$$

we look for a simpler approximate expression valid when $0 < a \ll 1$ and $m \ll n$, with n large.

- The first factor $n!/m!(n-m)!$ reads

$$\frac{1 \times 2 \times \cdots \times (n-m-1)(n-m)(n-m+1) \times \cdots \times (n-1)n}{m!(n-m)!}.$$

The numerator contains, first a factor $(n-m)!$, and then a product of m terms all practically equal to n (because $m \ll n$). Accordingly, simplification yields

$$\frac{n!}{m!(n-m)!} \approx \frac{n^m}{m!}.$$

- Since $a + b = 1$, the second factor reads

$$a^m b^{n-m} = a^m (1-a)^{n-m}.$$

The same condition $m \ll n$ leads us to replace the exponent $n - m$ by n, whence

$$a^m b^{n-m} \approx a^m (1-a)^n.$$

By virtue of $a \ll 1$, we can write, accurately to first order,

$$(1-a)^n = \exp[n \ln(1-a)] \approx \exp(-na).$$

Thus we are left with $a^m b^{n-m} \approx a^m \exp(-na)$.

- Let us now combine the two factors as transformed by our approximations:

$$P(m) \approx \frac{n^m a^m \exp(-na)}{m!} = \frac{(na)^m \exp(-na)}{m!}.$$

The product na we recognize immediately as the mean value $\bar{m}$ of the distribution. Accordingly, the end-result reads

$$P(m) \approx \frac{(\bar{m})^m \exp(-\bar{m})}{m!}.$$

This is the Poisson distribution; it depends on only the one parameter $\bar{m}$.

To determine its standard deviation, one starts from the expression stemming from the binomial distribution, $\sigma = (nab)^{1/2}$. Here $na = \bar{m}$, and $b = 1 - a \approx 1$, because a is very small. Hence $\sigma = (\bar{m})^{1/2}$.

4 The derivative of $\ln(m!)$ when m is large

When m is very large, a change of ± 1 may be considered as infinitesimally small on a scale of m. Even though m is a discrete variable, it can then be considered as if it were continuous, with increments Δx equal to 1. Hence we can write

$$\frac{d(\ln m!)}{dm} = \frac{\ln(m+1)! - \ln m!}{1} = \ln \frac{(m+1)!}{m!} = \ln(m+1).$$

Then, because 1 may be neglected compared to m, one has

$$\frac{d(\ln m!)}{dm} = \ln m.$$

5 The Gaussian (or normal) distribution

This is an approximation to the binomial distribution for very large values of n. Under this condition m, also very large, can be treated as a continuous variable, as we have just seen.

The binomial distribution reads

$$P(m) = \frac{n!}{m!(n-m)!} a^m b^{n-m}.$$

Take its logarithm:

$$\ln P(m) = \ln n! - \ln m! - \ln(n-m)! + m \ln a + (n-m) \ln b.$$

- We start by looking for the maximum of $\ln P(m)$, which occurs where

$$\frac{d}{dm} \ln P(m) = 0.$$

Differentiation yields

$$\frac{d}{dm} \ln P(m) = 0 - \frac{d}{dm} \ln m! - \frac{d}{dm} \ln(n-m)! + \ln a - \ln b.$$

Noting the approximation established in Appendix 4, we may write

$$\frac{d}{dm} \ln P(m) = -\ln m + \ln(n-m) + \ln a - \ln b = \ln \left(\frac{n-m}{m} \frac{a}{b} \right).$$

This derivative vanishes when the argument of the logarithm is equal to 1, a condition that reads

$$\frac{n-m}{m} \frac{a}{b} = 1.$$

In view of $a + b = 1$, the condition is met when $m = na$. At this value of m, $\ln P(m)$, and therefore $P(m)$, attain their maxima. Accordingly, na is the most probable value of m; but, as the binomial distribution shows, it is also the average value $\bar{m}$ of m. This coincidence is characteristic of symmetric distributions.

We stress the following results:

$$\frac{d}{dm} \ln P(m) = \ln \left(\frac{n-m}{m} \frac{a}{b} \right) \quad \text{and} \quad \frac{d}{dm} \ln P(\bar{m}) = 0.$$

- We shall now expand $\ln P(m)$ around the characteristic value $\bar{m} = na$, keeping terms only up to second order:

$$\ln P(m) \approx \ln P(\bar{m}) + \frac{d}{dm} [\ln P(\bar{m})] (m - \bar{m}) + \tfrac{1}{2} \frac{d^2}{dm^2} [\ln P(\bar{m})] (m - \bar{m})^2.$$

The calculation just above shows that the first derivative vanishes at $m = \bar{m}$. Hence we proceed directly to calculate the second-order term.

From

$$\frac{d}{dm} \ln P(m) = -\ln m + \ln(n-m) + \ln a - \ln b,$$

it follows that

$$\frac{d^2}{dm^2} \ln P(m) = -\frac{1}{m} - \frac{1}{n-m} = -\frac{n}{m(n-m)}.$$

For $\bar{m} = na$, and in view of $a + b = 1$, we have

$$\frac{d^2}{dm^2} \ln P(\bar{m}) = -\frac{1}{nab};$$

we recognize $nab = \sigma^2$, where σ is the standard deviation of the binomial distribution.

Accordingly, our expansion takes the simple form

$$\ln P(m) = \ln P(\bar{m}) - \frac{(m-\bar{m})^2}{2\sigma^2},$$

whence the distribution reads

$$P(m) = P(\bar{m}) \exp\left(-\frac{(m-\bar{m})^2}{2\sigma^2}\right).$$

- Next we must calculate $P(\bar{m})$, by exploiting the norming condition

$$\sum_{m=0}^{n} P(m) = 1.$$

Since m is being treated as a continuous variable, the (discrete) summation $\sum_{m=0}^{n}$ may be replaced by the (continuous) integral $\int_0^n$; in other words

$$\int_0^n P(m) \, dm = 1.$$

Because $P(m)$ is appreciable only in a narrow range of m-values around $\bar{m}$, the integration may be extended to infinity, on the understanding that $P(m) = 0$ when $m < 0$ or $m > n$. Hence

$$\int_{-\infty}^{\infty} P(m) \, dm = 1,$$

or equivalently

$$\int_{-\infty}^{+\infty} P(\bar{m}) \exp\left(-\frac{(m-\bar{m})^2}{2\sigma_2}\right) dm = 1.$$

We simplify the exponential by changing variables to $u = (m - \bar{m})/\sigma\sqrt{2}$; accordingly

$$P(\bar{m})\,(\sigma\sqrt{2})\int_{-\infty}^{+\infty} \exp(-u^2)\,\mathrm{d}u = 1.$$

Here we encounter a classic (and 'improper') integral, whose value is $\sqrt{\pi}$, and we conclude that

$$P(\bar{m}) = \frac{1}{\sigma(2\pi)^{1/2}}.$$

Thus the distribution law reads

$$P(m) = \frac{1}{\sigma(2\pi)^{1/2}} \exp\left(-\frac{(m-\bar{m})^2}{2\sigma^2}\right).$$

• It remains only to make the transition from the integer (hence discontinuous) variable m to a continuous variable x. To this end we choose an interval $[x, x+\Delta x]$ such that $1 \ll \Delta x \ll m$. This is perfectly possible because m is very large (Fig. 2.12).

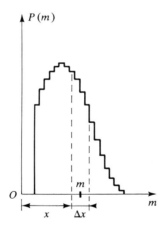

Fig. 2.12 Passage from the discrete variable m to the continuous variable x.

The number of values of m in this interval is Δx (within ± 1), all of them such that $x < m < x + \Delta x$.

The probability of finding m in this interval is $P(m)\Delta x$ (as a consequence of the theorem of total probabilities).

To emphasize the passage to a continuous variable, we write this probability as $\rho(x)\Delta x$; it is given by

$$\rho(x)\Delta x = \frac{1}{\sigma(2\pi)^{1/2}} \exp\left(-\frac{(x-\bar{x})^2}{2\sigma^2}\right)\Delta x,$$

where x is a continuous random variable (equal to m when x is an integer), and $\rho(x)$ is the probability density (equal to $P(m)$ when x is an integer).

To summarize: we have derived the Gaussian (normal) distribution for a continuous variable x; it reads

$$\rho(x) = \frac{1}{\sigma(2\pi)^{1/2}} \exp\left(-\frac{(x-\bar{x})^2}{2\sigma^2}\right),$$

where $\rho(x)\Delta x$ is the probability of finding x in the interval $[x, x+\Delta x]$, $\bar{x}$ is the average and also the most probable value of x, and σ is the standard deviation.

Accordingly, this distribution depends on two parameters, $\bar{x}$ and σ.

Further reading

See Reif (1967) in the Bibliography.

Karl Friedrich Gauss. German astronomer, mathematician, and physicist; born in Brunswick, died in Göttingen (1777–1855). All his working life is spent in Göttingen: he is director of the observatory as well as professor of astronomy. His contributions to the exact sciences are enormous, and extend over all its branches: to this day we teach Gaussian approximation, Gauss's theorem, Gauss's principal points, and the Gaussian distribution. It is this last, together with his discovery of the method of least squares, that enables Gauss to propose a theory of physical measurement. (Palais de la Découverte)

3

Gaussian deviations (the theory of physical measurements)

Humanum fuit errare, diabolicum est per animositatem in
errore manere . . .†
St. Augustine

Though it is the quantitative science par excellence, physics does not in the end deserve to be described as a wholly exact science. Between the true and the measured values of any quantity there is always some difference, which our remote ancestors chose to call an 'error'. It is this choice of word, as unfortunate as they come, that has made generations of physicists feel guilty, and it is their feelings of guilt that have given rise to the traditional calculus of errors. This calculus is not without merit as regards a single measurement of a continuously variable quantity; but it is over-pessimistic as regards repeated measurements of such a quantity, because it disregards the increased precision available from accumulating experimental results. Moreover, it has nothing to say about a fundamental feature of modern measurements, namely the counting of particles, where the quantity measured is discontinuous. Fortunately, in the last several years the fog has been lifting, and both in schools and in universities one does now find approaches to the problem that are more realistic and more in tune with the experimental methods actually used in research. At long last we have stopped applying the word 'error' to a difference that is perfectly normal. Our intended small revolution in teaching must therefore start with a revolution as regards the meaning of words, not fully accomplished as yet, but one to which we wish to make our own modest contribution. Accordingly, we adopt the following designations.

- A *random deviation* is what used to be called an 'accidental error'; it corresponds to the random features of the measurement process.

- A *systematic shift* is what used to be called a 'systematic error'; it corresponds to the deterministic features of the measurement process.

- *Error* is anything due to departures from the proper rules of measurement; it

† To err was human; but to persevere in error from sheer obstinacy is diabolical.

includes human error, like misreading a scale, and also mechanical errors like the malfunctioning of an amplifier.

As to the word 'uncertainty', which physicists have a habit of applying to the standard deviation, we do not use it at all, because it is overburdened with tacit implications as a result of its very debatable employment in quantum mechanics (Section 7.6).

Finally, we must stress that our point of view is that of a teaching or of a research laboratory, and not that of a metrological laboratory. The latter is in the business of devising very precise standards, which, for us, constitute the true value of the physical quantity we wish to measure. Our aim is to assign to every result of an ordinary measurement a certain confidence interval, chosen so that we can accept, at some predetermined level of confidence, that the true value lies within the interval.

3.1 The central limit theorem

The universal nature of the Gaussian distribution could hardly fail to engage the attention of mathematicians, who then embarked on a dogged pursuit of the deep reasons for it. This led them to prove, step by step, an absolutely basic theorem, the so-called *central limit theorem*.† Our parable of the average holidaymaker choosing his spot on the beach was an attempt to make the reader discover the intuitive essence of the theorem from scratch; summarized rather tersely it might read 'whatever you sum you get a Gaussian'. The following is a scientifically somewhat more conventional formulation.

If an overall random variable is the sum of very many elementary random variables, each having its own arbitrary distribution law, but all of them being small, then the distribution of the overall random variable is Gaussian.

The scope of this theorem is very wide, and we proceed to use it to analyse the typical measuring process.

Though very simple in principle, an ordinary length measurement will serve to exhibit all the essentials of such a process.

On a horizontal table covered in plastic, we use a fibre-tip pen to draw two indelible marks perpendicular to an edge. We check by eye that the table is flat and the edge straight. Our aim is to measure the distance between the marks, initially by using a rather primitive ruler made from a 10 cm strip of millimetre graph-paper. The measurement consists in successive displacements of the ruler AB, taking great care as we slide it along to align the ith position of end A

---

† Though its application is not totally unrestricted, the theorem has been proved, under successively less stringent conditions, by Gauss, by Liapunov, and by Lindeberg.

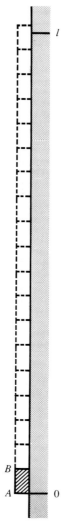

Fig. 3.1 Length measurement with the ruler *AB*
(10 cm piece of millimetre graph-paper).

with the $(i-1)$th position of end *B*. To ensure this, we make marks on the plastic cover with a very thin fibre-tip. After thus shifting the strip 19 times (which calls for a joint total of 38 operations of marking and alignment), we obtain the result 1861 mm. To start the same process again, we first rub out all the marks with methylated spirits, to prevent them from influencing the subsequent results. Though all the measurements are made by the same operator, this precaution is absolutely essential. In this way we complete 15 measurements in only two hours; this is short enough to consider the humidity of the air as constant, whence the changes in the length of the strip of graph-paper are negligible. The results are as follows:

1859 mm	once
1860 mm	once
1861 mm	4 times
1862 mm	6 times
1863 mm	3 times.

They are entered on a histogram (the left-hand side of Fig. 3.2a), which provides ample evidence that the results are scattered. What is the reason for the scatter? Essentially it lies in the 38 operations needed for every measurement. However careful one is, these are subject to arbitrary deviations in one direction or in the other. The overall deviation, the sum of 38 small random deviations, is appreciable, and its distribution is Gaussian. This is what the central limit theorem tells us, and this is what is suggested (albeit in discretized fashion) by the histogram displaying our set of 15 results.

It remains only to compare our set of homespun measurements with a more accurate set obtained with the usual two-metre steel tape graduated in millimetres. In this case one needs only two operations, and the result reads 1868 mm. Repeating the operation 15 times as before we get the same value of 1868 mm every time; graphical representation is therefore very simple (as on the right of Fig. 3.2a). The uniqueness of the result stems from the fact that only two decisions are involved. Random deviations are much smaller, and are confined to the interval 1868 ± 0.5 mm, so that the end-result always reads 1868 mm.

We have now acquired all the ingredients needed for constructing a mathematical model for these two series of measurements. Call the probability density $\rho(x)$; hence the probability that the result of a measurement lies in the range $[x, x + \Delta x]$ is $\rho(x)\Delta x$. The two experimentally determined histograms yield, for $\rho(x)$, the two distributions shown in Fig. 3.2b.

- For the first set of measurements (on the left of Fig. 3.2b), made with the graph-paper, the distribution is Gaussian; this shape is suggested by the histogram and conforms with the central limit theorem. The mean value of the distribution is $x = 1861.6$ mm, and its standard deviation is $\sigma = 1.1$ mm.†

- In principle, the second set of measurements (made with the steel tape) should likewise be modelled by a Gaussian distribution, but by a much narrower one. Its mean value would be $l = 1868$ mm, and its standard deviation would be much smaller than 1.1 mm. This leads us to the perfectly

† The numerical method used here is explained in Section 3.3.

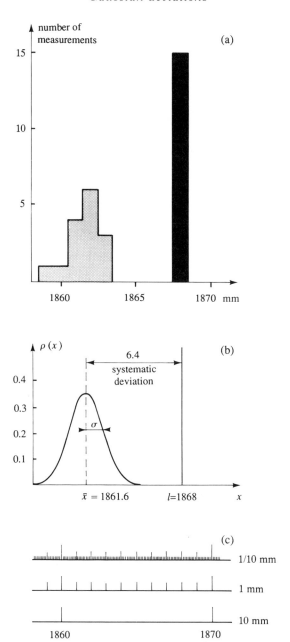

Fig. 3.2 (a) Histogram of 15 measurements of the same length: left, with a 10 cm piece of millimetre graph-paper; right, with a two-metre steel tape. (b) The corresponding probability distributions: left, a Gaussian; right, the result is unique. (c) Three scales, defining three different levels of sensitivity: 0.1 mm, 1 mm, 10 mm.

well-warranted approximation of replacing such a narrow Gaussian by its limit as the standard deviation tends to zero. The limit is merely an infinitely high vertical line drawn upwards from the point $l = 1868$ mm of the horizontal axis (see the right of Fig. 3.2b). From this point on, we assert that, as regards the first set of measurements, $l = 1868$ mm represents *the true value of the quantity being measured*, on the grounds that the second set of measurements is much more accurate than the first.

We can now proceed to analyse the differences between the results of the first set of measurements and the true value of l; this will allow us to introduce the three most basic characteristics of any measurement process.

3.2 Reproducibility, bias, and sensitivity

The *relative deviation* is a parameter related to the *reproducibility* of a measurement. We have just shown that every operation introduces some random deviations, beyond the control of the operator. These are well characterized by the standard deviation σ, and better still by the relative deviation $\sigma/\bar{x}$, where $\bar{x}$ is the average of the measured values. A measurement is the more closely reproducible, the smaller the relative deviation $\sigma/\bar{x}$. Some feeling for this can be gained from the following orders of magnitude of relative deviations:

mainly qualitative demonstration experiment	10^{-1}
routine measurement	10^{-2}
research work	10^{-3} to 10^{-4}
metrology	10^{-6} to 10^{-10}.

Note that we have talked of reproducibility as a property of a measurement process rather than of the apparatus. In using our strip of graph-paper for instance, one must expect the quality of the results to depend on the experimenter even if all the experimenters are careful, simply because they do not all have the same acuity of vision and the same manual dexterity. Accordingly, the standard deviation is a characteristic of physicist plus apparatus jointly. This does not of course preclude two operators from obtaining the same result, but we do know that their confidence intervals will differ. However, modern measurement techniques are largely immune to such difficulties, because they tend to use detectors producing electrical signals that are recorded and analysed automatically, side-stepping personal input from the operator.

Absence of bias† relates to the freedom of the results from systematic deviations. Our first set of length measurements (Fig. 3.2b) certainly does suffer from such deviations. The average of the 15 measurements is 1861.6 mm. The deviation from the true value, expressed in millimetres, averages $1868 - 1861.6 = 6.4$. It stems from the length of our paper ruler, which is greater than 100 mm, the relative excess as revealed by the experiment being $1868/1861.6 = 1.003\,438$. Here we are evidently faced with something perfectly deterministic. In order to pass as unbiased, every raw measurement must first be corrected so as to eliminate systematic deviations. In our present case, this correction consists in multiplication by the factor 1.003 438.‡ Here the relative systematic deviation is $6.4/1861.6 = 3.4 \times 10^{-3}$. The smaller the relative systematic deviation, the less biased we say is the measuring process.

Our experience with measuring lengths indicates the procedure to be followed in order to diminish bias. To validate an ongoing series of measurements of some given quantity, one must first calibrate them by means of a set of accurate measurements of the same quantity. From these, one determines a *correction term* (or *correction factor*), which is then applied to all the subsequent measurements of the quantity in question. This is what we shall do in the next section.

Note finally that, in modern electrical measuring apparatus, the preliminary calibration is embodied in the guarantee given by the manufacturer. When a voltmeter is said to be of grade 2, this means that the systematic deviation does not exceed 2% at end-range. It may not come amiss to check the calibration, especially if the voltmeter is not of the newest.

Sensitivity is a property related to the graduations of the scale on which the apparatus registers the results. In our length measurements, the sensitivity is governed by the spacing between successive subdivisions, which is 1 mm. In general terms, *the sensitivity of the apparatus is the smallest measurable difference between two different values of the variable*; this difference might correspond to the width of one division on a screen, or to the value of the smallest unit if the meter is digital. It is worth noting in this context that, even when measuring a continuous variable, the raw result always varies discontinuously.

Taken by itself, our rather simplistic definition is not enough to determine the sensitivity appropriate to every measurement. To elucidate this point, we revert to our elementary example of measuring lengths, and ask ourselves two questions (Fig. 3.2c).

1. What would happen if we used graph-paper divided into centimetres,

† Translator's note: we avoid the word 'accuracy', which in English has no commonly accepted and well-defined meaning, while 'unbiasedness', though sometimes encountered, is barbaric.

‡ It is essential that the correction factor be given at least to two significant figures beyond the result to be corrected. We shall discuss this point in the next section.

that is if the width of the smallest subdivision were increased tenfold? Before correction, every measurement then gives the same raw result, namely 186 cm, which implies a result between 185.5 and 186.5 cm. All random aspects of the measurement process have been obliterated. The precision of the result is poor, since it will have to be given as 186 ± 0.5 cm instead of 1861.6 ± 1.1 mm. By choosing too wide an interval we have lost some of the profits available from the measurement process.

2. What would happen if we stuck to our homespun method but used a Vernier scale calibrated to 0.1 mm? The process would take longer and would become more delicate, but this would make no difference to the accuracy of the 38 operations, and the scatter of the 15 raw results would therefore remain much the same. The 15 values would merely be quoted to five instead of four figures, which is pointless. By choosing too narrow an interval, we have made the process more awkward without improving the results.

Accordingly, every measurement process has an *optimal range of sensitivity*. Changing the sensitivity by an order of magnitude, that is raising or lowering it by a factor of 10, is certainly to be avoided; the most one might envisage are some manoeuvres confined to half an order of magnitude, that is within a factor of three. But just what is the optimal sensitivity? The answer to this question is suggested by our experience with measuring lengths: the graduations are spaced by 1 mm, and the standard deviation is 1.1 mm. Accordingly it is sensible to compromise by choosing the sensitivity equal to the standard deviation σ. If the graduations are spaced by more than 3σ, the sensitivity is too low; if the spacing is less than $\frac{1}{3}\sigma$, the sensitivity is excessive; the optimum sensitivity is near σ.

In fact, makers of measuring apparatus adopt a somewhat more subtle compromise to obtain the best ratio of quality to price. Given the price, they optimize three parameters, namely relative deviation, bias, and sensitivity, by means of an analysis of the same kind as we have just given. The performance achieved is indicated on the label on the apparatus, which guarantees its calibration and the specified precision; both result from technical compromises between the three often conflicting requirements of reproducibility, absence of bias, and sensitivity.

3.3 The measurement of a continuous variable

With preliminary calibration and the choice of the optimum sensitivity behind us, we revert to our millimetre graph-paper, and set about measuring, on the same horizontal table, the separation between two newly made benchmarks.

It is perfectly possible to settle for just one measurement. After 17 displacements of the ruler, corresponding to 34 operations, we obtain 1627 mm as our raw result. We know that it is subject both to random and to systematic deviation.

• As regards the random deviation, we would like a realistic estimate of it; but we have absolutely no idea whether the elementary deviations have, by accident, cancelled out, or whether, unfortunately, they have all combined in the same direction. All we can do is to set an upper limit corresponding to the worst case. Provided we have worked carefully, we can guarantee that in each operation the random deviation cannot exceed 0.25 mm. Hence the overall random deviation cannot exceed $34 \times 0.25 = 8.5$ mm.

• As regards systematic deviations, we can exploit the correction factor of 1.003 438 available from our earlier calibration (Section 3.2); this can be used to adjust both the raw result (1627 mm) and the random deviation (8.5 mm). Accordingly we obtain

$$1.003\,438 \times 1627 = 1632.5936 \quad \text{and} \quad 1.003\,438 \times 8.5 = 8.5292.$$

Therefore the end-result reads $x_0 = 1632.6 \pm 8.5$ mm.

The relative deviation is 5×10^{-3}, which is respectable but no more. The quality of this result is limited by the fact that, in order to be on the safe side, we have deliberately allowed for the least probable outcome, greatly exceeding thereby the most probable value of the random deviation.

Repetition of course takes longer, but it gives a much better result. The same operator now uses the same graph-paper ruler to make 15 measurements, each calling for 34 alignments. Here are the results:

1627 mm	3 times
1628 mm	7 times
1629 mm	4 times
1631 mm	once.

The total number of measurements is $n = 15$.

As was to be expected, we are now faced with a random process, and should like to know the mean value $\bar{x}$ and the standard deviation σ. The true values of these quantities are forever beyond our ken, but probability theory will supply realistic estimates of them, which we shall denote by $\bar{x}^*$ and σ^*. Accordingly, attention now focuses on conclusions drawn from the theory.

• The estimate $\bar{x}^*$ of the mean $\bar{x}$ is the arithmetic mean of the n experimental results. Thus

$$\bar{x}^* = \frac{1}{n} \sum_i x_i.$$

- The estimate σ^* of the standard deviation σ is

$$\sigma^* = \left(\frac{\sum_i (x_i - \bar{x}^*)^2}{n-1} \right)^{1/2}.$$

Both relations stem from properties of the Gaussian distribution.

- The numerical evaluation is readily performed on a pocket calculator (see Appendix 1), and we find $\bar{x}^* = 1628.2667$ and $\sigma^* = 1.0328$.

- We are now in a position to define and to estimate the *confidence interval* and the *confidence level*. To this end, we need Student's distribution for $n-1$ degrees of freedom (see Appendix 2). Discovered shortly before the Second World War, this distribution remained secret until 1945, because of its exceptional reliability in the quality control of production lines manufacturing arms and projectiles. Here we shall exploit it more pacifically. For a set of n measurements, we denote the confidence level by β and the corresponding safety factor by t_β. From Student's theory, we then learn that

$$\text{Confidence interval} = \pm t_\beta \sigma^*/n^{1/2},$$
$$\text{Confidence level} \quad = \beta.$$

In other words, the theory tells us that the probability for the result of our measurement of x to lie in the range $\bar{x}^* \pm t_\beta \sigma^*/n^{1/2}$ is equal to β. The theory is tabulated (see Appendix 2), and gives us the value of t_β for each pair of values (n, β). For numerical application to our set of measurements, we put $n=15$ and $\beta=0.95$. Student's tabulation then yields $t_\beta=2.15$, and the end-result of the calculation is a confidence interval of ± 0.5733.

- Applying the correction factor is the last but one step in our procedure, and eliminates systematic shifts from the mean and from the confidence interval. Accordingly, we now multiply the results by $1.003\,438$ and find

$$1.003\,438 \times 1628.2667 = 1633.8647 \quad \text{and} \quad 1.003\,438 \times 0.5733 = 0.5752.$$

How many of these digits should we retain?

- Choosing the number of significant figures is the last step in the procedure. The calculations so far have been carried through with at least two more figures than signify. This forestalls any errors (in the strict sense of the word) that the calculation itself might introduce into the end-result. The advent of pocket calculators has made this very easy, as the present generation knows only too well. The next decision by contrast is rather more delicate and requires some modicum of physical insight.

At the 0.95 confidence level we express the confidence interval to just one significant figure:† 0.5752, or rather 0.6. It is this last result that dictates our attitude to the mean: 1633.8647, or rather 1633.9. The first four digits are significant, while the fifth specifies the centre of the confidence interval. Hence the end-result reads

$$x_0 = 1633.9 \pm 0.6 \text{ mm at a confidence level of } 0.95.$$

It is particularly informative to compare our initial single measurement with the repeated measurements. We recall the results:

single measurement: $x_0 = 1632.6 \pm 8.5$ mm
repeated measurements: $x_0 = 1633.9 \pm 0.6$ mm.

The first thing to note is that, encouragingly, the two results are compatible. More important, the second value is evidently more precise than the first by at least one significant figure, so that the extra effort does indeed produce an improvement.

The method of repeated measurements fits very easily into the usual pattern of practical work. All that is needed is to make all the students measure the same physical quantity (density, refractive index, wavelength, etc.) with the same apparatus under the same conditions, collate the results, and then through Student's distribution obtain an end-result to excellent precision. It can happen on such occasions that the overall harmony of the proceedings is troubled by one result in clear disagreement with the rest. Should one keep it nevertheless? Or should one rather discard it in the belief that the aberrant result stems from an error of the kind described near the start of this chapter, i.e. from some failure human or material? To decide this question we need some plausible selection criterion.

The *criterion for observational error* stems directly from the geometrical properties of Gaussians. The interval $\pm 3\sigma$ corresponds to a probability of 0.997. Hence the probability for a single measurement to lie outside this interval is 0.003, which is very small. The criterion applies to the raw data, before any corrections are applied. For example, of our 15 length measurements, the one value of 1631 mm is quite far from the rest, and we should ask ourselves whether it is erroneous. The mean before correction is 1628.3 mm, and the standard deviation 1.03 mm. Hence the interval $\pm 3\sigma$ stretches from 1625.2 to 1631.4 mm. The value of 1631 mm lies within the interval, and should not be discarded. By contrast, if we had found a value of 1633 mm, we should have discarded it, and should then have had to repeat the calculations with the remaining 14 instead of the original 15 results. The criterion $\pm 3\sigma$ is not an absolute (and nor are any other probabilistic criteria); but the

---

† At the 0.99 confidence level, one would be better advised to give the interval to two significant figures.

associated confidence level is high enough to allow one to suppress an aberrant measurement, without being accused of doctoring the results.

The *quality of a set of data* is a criterion one applies *a posteriori*, in order to decide whether the work has been satisfactory. In following the procedures we have indicated, preliminary calibration and adoption of the optimal sensitivity are taken for granted. Hence it is the random features that dominate the operation, and they are characterized by the parameter $t_\beta\sigma^*/\bar{x}^*n^{1/2}$ or in other words by *half the relative confidence interval*. A shorter designation would be desirable, but would plunge us back into the morass of semantics. We might speak of 'imprecision': the better the measurement, the lower the imprecision. This would be quite logical but it hardly trips off the tongue.

It makes more sense to consider the inverse of this ratio, namely $x^*n^{1/2}/t_\beta\sigma^*$, which we shall call the *precision*: the better the measurement, the higher the precision. For our set of 15 measurements of one and the same length, the precision is 2.7×10^3. For a 10 cm strip of millimetre graph-paper this is not too bad at all!

In the last analysis it is their respective precisions that allow one to compare two different sets of data, though of course only on condition that both are expressed in terms of the same confidence level β.

3.4 The measurement of a discontinuous variable

Particle counting is a very common practice in modern physics laboratories. That is what one does, for instance, to determine the activity of a radioactive source (see Fig. 3.3).

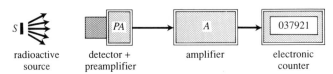

| radioactive source | detector + preamplifier | amplifier | electronic counter |

Fig. 3.3 Sketch of a counter used in nuclear spectroscopy.

The radioactive source emits ionizing particles. Those that are stopped in the detector produce an electrical signal, which after amplification can trigger an electronic counter. The result of the measurement is the number m of counts in a time interval $t_1 - t_0$; in the jargon of the trade m is sometimes called 'the number of clicks', recalling the time when it was the physicist's ear that served as the counter.

Thus, in one particular experiment, we counted m = 102 *events over a period of 15 minutes. What can we make of this result?*

The number m is *exact*. This means that it is exactly equal to the number of times the detector has been triggered in the time interval $t_1 - t_0$. But if we repeat the measurement with an equally long interval $t_2 - t_1$, we shall obtain a different value m', which is also exact. How come the difference? Simply because we are engaged in recording a random process, as we can tell for three different reasons.

1. First, radioactive decay itself is a random process.

2. Next, the stopping of the particles by the detector is also a random process. However, for α and β particles, the stopping probability (called the detector efficiency) is very close to 1 (i.e. to certainty). By contrast, for γ rays it is at most 0.1; hence it is especially for the latter that the random nature of the detection process comes into play.

3. Finally, and less important, we should take into account the random features in the operation of the electronic counter, which cannot ensure that the time intervals $t_1 - t_0$, $t_2 - t_1$, $t_3 - t_2$ are exactly equal.

It follows that the number m of counts, exact but random, cannot directly represent the result of the measurement. What we need is an estimate $\bar{m}^*$ of the mean number $\bar{m}$ of events, and it is the number $\bar{m}$ that embodies the result.

The solution comes from probability theory. For every particle that is emitted there are two possibilities: detection, with low probability, and non-detection, with high probability. The process falls within the scope of the binomial distribution; moreover, the Poisson approximation is valid because the average value $\bar{m}$ is not very large. Hence we know the distribution law (the probability distribution and the standard deviation) for the m:

$$P(m) = \frac{(\bar{m})^m \exp(-\bar{m})}{m!} \quad \text{and} \quad \sigma = (\bar{m})^{1/2}.$$

Our method hinges on the assumption that, though $\bar{m}$ may not be very large, neither is it too small. Thus, for $\bar{m} > 50$, the geometrical features of the Poisson distribution are to a very good approximation the same as those of a Gaussian.[†] For instance, we might assert that the number m of counts lies in the confidence interval $\bar{m} \pm \sigma$ with a confidence level of 0.68. Then, if only we knew $\bar{m}$ and $\sigma = (\bar{m})^{1/2}$, we could make a probabilistic prediction about m.

---

† We are dealing with a discontinuous Gaussian distribution, i.e. with a distribution whose variable assumes only integer values.

But what the measurement actually gives us is m; it is $\bar{m}$ and σ whose values we lack and wish to estimate. To this end, we make two observations.

1. The probability of finding m in the interval $\bar{m} \pm \sigma$ is the same as the probability of finding $\bar{m}$ in the interval $m \pm \sigma$. In fact this probability corresponds to a distance of at most σ between m and $\bar{m}$, and it is quite

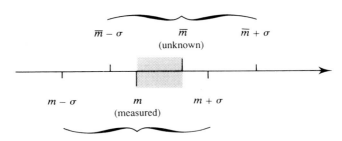

Fig. 3.4 How to go from measured m to unknown $\bar{m}$.

irrelevant whether one measures the distance from m or from $\bar{m}$. It is represented by the hatched segment in Fig. 3.4. This leads us to choose the measured value m as our estimate $\bar{m}^*$ of the unknown mean value $\bar{m}$. It remains only to estimate the confidence interval.

2. The standard deviation must be found from the relation $\sigma = (\bar{m})^{1/2}$, but here again we are ignorant of $\bar{m}$. However, subject to the condition $m > 50$, the difference between m and $\bar{m}$ is small relative to the value of either ($< 14\%$), because it is of the same order as the standard deviation. *A fortiori*, the relative difference between σ and σ^* will be smaller still ($< 7\%$), if we replace $\sigma = (\bar{m})^{1/2}$ by the estimate $\sigma^* = (\bar{m}^*)^{1/2}$, where $\bar{m}^* = m$. Hence $\sigma^* = m^{1/2}$ is a good estimate of σ. As for the confidence level, corresponding to one standard deviation it is 0.68.

To summarize, we adopt the following rule for determining the unknown average value $\bar{m}$.

- Our estimate of $\bar{m}$ is the actual number of counts m.
- Our estimate of σ is $m^{1/2}$.
- The result reads: $\bar{m} = m \pm m^{1/2}$ at a confidence level of 0.68.

One could of course raise the confidence level by choosing a wider confidence interval. For instance, the interval $\pm 2m^{1/2}$ corresponds to the level 0.95, and the interval $\pm 3m^{1/2}$ to 0.997.

The duration of the count is one key element in the operation. In the example we chose, the number of counts, in a period of 15 minutes, is $m = 102$. Hence the result reads: $\bar{m} = 102 \pm \sqrt{102}$, or in other words $\bar{m} = 102 \pm 10$ at a confidence level of 0.68.† The relative deviation $\sigma/\bar{m}$ here is 10%. This is a rather modest result; one might well ask oneself what one would need to do in order to improve the precision by a factor of 10, that is to lower the relative deviation to 1%. The obvious idea is to extend the duration of the measurement, but this is very time-consuming. The relation

$$\sigma/\bar{m} = (\bar{m})^{1/2}/\bar{m}$$

shows that the number of counts would have to be increased by a factor of 100; hence the duration would have to be extended by the same factor, from 15 minutes to 25 hours. Is this in fact a reasonable proposition? Over so long a time, could one prevent the apparatus from drifting? Of course, everything depends on the quality of the equipment. If it is adequately stabilized, then such a long-drawn-out measurement does become possible; but then it has to be automated, and it becomes essential to print out the partial results at regular intervals (say every hour), so that one can check afterwards whether the system was operating normally throughout.

In this respect the teacher is usually better placed than the researcher. Experiments set up for students produce enough counts for the measurement to yield reasonable results in the time span of a standard session in the laboratory.

In order to proceed from m to A, that is from the number of particle counts to the activity of the source, one uses the relation

$$m = A \, \frac{\Omega}{4\pi} \, K(1 - F) \, (t_1 - t_0),$$

where Ω is the solid angle subtended by the detector at the source, K is the efficiency of the detector, that is the probability that it will stop a given ionizing particle, $t_1 - t_0$ is the duration of the count, and F is the loss rate of the counting circuit. The losses are due to fortuitous coincidences between events; the problem of calculating F is the same as the problem of the jammed phone line, both being governed by the Poisson distribution.‡

---

† It might seem more precise to write $\sqrt{102} \approx 10.1$, but with the confidence level at 0.68 such precision is illusory; moreover, for $m = 102$ there is still a difference of the order of 1% between the Poisson and the Gaussian distributions. Note also that the true (and unknown) value of m is very likely fractional, while its estimated value, namely 102, is an integer.

‡ The factor $1 - F$ is nothing but the probability $P(0)$ in the Poisson distribution. Hence one has

$$1 - F = \exp\left(-\frac{m\tau}{t_1 - t_0}\right),$$

where τ is the duration of a single pulse. If the exponent is much smaller than 1, then one can expand to first order, which eventually leads to

$$F \approx \frac{m\tau}{t_1 - t_0}.$$

The significance of F is now clear: it is the combined duration of the actual counts divided by the total observation period.

The activity A is expressed in becquerels (1 becquerel = 1 decay per second).

The determination of each of these terms is a complicated matter, calling for preliminary measurements and calculations. However, given a well-calibrated source, one can determine the product of all the four factors multiplying A at one go. Here one must be particularly careful, because it is with these measurements that systematic deviations will creep in. They can never be eliminated completely, but it is the job of the experimenter to try and reduce them below $\frac{1}{3}\sigma$, so as to prevent them from exerting any significant influence on the random deviations.

We note in conclusion that the measurement of the activity of a source is by no means exempt from the general rule: it has random features, represented by the confidence interval, and deterministic features, represented by systematic deviations that bear on the factors that multiply A. The latter persist, unchanged, even if the experiment is repeated.

3.5 The smoothing of experimental data

Smoothing is a very important step on the long road that leads the physicist from experiment to law of nature. Consider as an example the following data (I, U) connecting the voltage U with the current I across a Leclanché cell, in an attempt to determine its electromotive force (EMF) and its internal resistance. Our immediate aim is to determine, as best we can, the relation $U = \phi(I)$.

The rule of thumb, popular ever since there have been physicists, is to rely on trained hand and trained eye *to draw the curve which runs the closest possible to all the points*. Naturally this assumes that the relation under study is regular in some sense, and in the ideally preferred case linear. This is the case in the example we have chosen, where we measure the EMF E and the internal resistance r of a cell by using the law $U = E - rI$. The results are shown in the table below, and plotted in Fig. 3.5.

U (V)	4.49	4.47	4.45	4.43	4.39	4.36	4.32	4.27	4.22	4.15	4.06	3.97
I (A)	0.010	0.020	0.030	0.040	0.060	0.080	0.100	0.130	0.160	0.200	0.250	0.300

The first check is to position one eye so that it looks along the line of points. This shows, first, that the points do indeed suggest a straight line, and, second, that none of the points is very far off this line. If any were, they would have to be considered as aberrant.

Having taken these precautions, we use a transparent ruler to draw the best line we can through the points. The EMF we seek is given by the intercept on

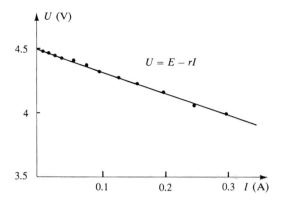

Fig. 3.5 Measurement of the internal resistance and EMF of a generator, using a graphical method.

the vertical axis, while the internal resistance is given by the (absolute value of the) slope. In this way we find that $E=4.50$ V and $r=1.80\ \Omega$.

The virtue of this rule of thumb is that it is simple and effective; on the other hand it lacks the elements that would allow the deviations to be determined correctly. In this particular case it would therefore be prudent to limit the display to two significant figures. If greater precision is required, then one must switch to a method that is more sophisticated and better founded mathematically.

The *method of least squares* is a smoothing technique based on the mathematics of Gaussian distributions. It enables one to find the most probable match between the theoretical law $y=\phi(x)$ and the set of experimental points (x_i, y_i). If, for example, the theory asserts that $y=ax+b$, then least-squares fitting allows one to determine the parameters a and b. The technique consists in minimizing the quantity

$$\sum_i (\text{experiment} - \text{theory})^2,$$

namely the sum of the squared differences between the measured values of y_i and the corresponding theoretical values $\phi(x_i)$. Two conditions must be met if this technique is to be readily applicable.

1. Only the y-coordinates may be subject to random deviations. By contrast, the x-coordinates are considered to be exact. In practice this means that x_i is measured far more precisely than y_i.†

---

† If one works with meters provided with a graduated scale (say ammeters and voltmeters), then one can choose a procedure that does satisfy this condition. First one adjusts the control parameter x in the circuit so that the ammeter needle coincides with one of the graduations; with sufficient care, the random deviation of the variable x does not exceed one-tenth of a scale division. Then one measures the corresponding response (the variable y), whose random deviation is necessarily just one division. Hence x is measured ten times more precisely than y.

2. All the data y_i have the same standard deviation σ.†

Once these two conditions are met, the calculation is not difficult; however, it is lengthy, especially if there are more than two adjustable parameters, and if the theoretical relationship fails to assume a simple mathematical form. On the other hand, if this relationship takes the form $y = ax + b$, then the calculation is much easier; it is called *linear regression*, and is programmed into most scientific calculators.

Let us revert to our measurement of E and r, and look for the parameters of the *regression line* using an HP-32E calculator (see Appendix 3). It yields three results:

- the y-coordinate at $x = 0$:   4.502 392;
- the slope: $-1.774\,421$;
- the correlation coefficient: $-0.999\,818$.

The first two values entail $E = 4.502$ V and $r = 1.77\ \Omega$. Moreover, probability theory supplies a method for estimating standard deviations that corresponds to the results from linear regression. The calculation is rather lengthy, and eventually yields the end-results $E = 4.502 \pm 0.004$ V and $r = 1.77 \pm 0.02\ \Omega$.

Notice that we have gained one significant figure beyond the rule-of-thumb, whose precision is actually quite difficult to estimate; but one should bear in mind that the rule-of-thumb does of course have the advantage of simplicity.

As to *the correlation coefficient*, it characterizes the *goodness of fit*. The fit would be perfect if the coefficient were -1 (negative slope) or $+1$ (positive slope). The value we find here is $-0.999\,818$, which is remarkably close to -1. It is reasonable to rate the fit as excellent, though one should realize that in fact the test is extremely sensitive. If the coefficient were only -0.99, it would already signal that the experimental points display some curvature visible to the naked eye, or at the very least that they are appreciably scattered.

This last remark raises the question of the choice of the theoretical relationship. We have used linear regression because *we know* that this relation reads $y = ax + b$, or else because *we choose* to restrict ourselves to displaying the phenomenon empirically by means of the simplest possible mathematical formula. The risk lies in the tempting simplicity of the appeal to the calculator. Even if we are faced, for example, by a cluster of points drawn on millimetre graph-paper by a monkey, there is nothing to stop us from entering these data into the calculator and from them determining the regression line. We always get a result, but its correlation coefficient would be very close to zero, and the result therefore very bad; and no interesting conclusions could be drawn from it. In other words, one must not misuse linearization procedures when the

---

† This condition is well enough satisfied if the operating procedure is the same for all the values of y_i.

correlation coefficient shows that they are suspect. What do we do when they are?

Even if our experimental data are more significant than monkey doodles, we must still process them through a least-squares fit; but we may need to use relationships whose mathematical form is more complicated (parabolic, exponential, sinusoidal, etc.). The form might be chosen arbitrarily in the course of a naïve exploration of possibilities, or it might correspond to some known theory of the phenomena under study. But in all cases one needs a criterion for assessing the goodness of the least-squares fit. Such a criterion exists, and is given by the χ^2 test due to Pearson.

The parameter χ^2 is defined by

$$\chi^2 = \sum_i \frac{(\text{experiment} - \text{theory})^2}{(\text{standard deviation})^2}.$$

Its optimal value equals the number of experimental points less the number of adjustable parameters. If χ^2 is significantly higher than the optimal value (by more than a factor of 10), then this is probably because the relation being tested is unsatisfactory. If, on the other hand, χ^2 is close to its optimal value, then the theoretical relation to which one is fitting is sensible, that is it has a high probability of being right. Though the χ^2 test cannot warrant a theory automatically, it does at least allow one to separate the wheat from the chaff.

These rather sophisticated mathematical methods are awkward to implement, unless one already has a program for putting them on to a microcomputer; that is certainly the way of the future. For the present, and for those who need more than the special case where $y = ax + b$, the next section describes a method that is simple and quite widely applicable, based on transformations of the mathematical relations.

3.6 Relations transformed

At the outset one has no idea at all about the mathematical form of the relation $y = \phi(x)$ between the two physical quantities y and x. But if we can transform it into a relation of the form $y = ax + b$, then it immediately becomes far easier both to verify and to use. Such transformations play a very important part in physics.

Classic examples are exponential decay laws of the type $y = y_0 \exp(-\alpha x)$, applicable for instance to

- radioactive decay, $N = N_0 \exp(-\lambda t)$;
- the barometer formula, $P = P_0 \exp(-mgz/kT)$;

- the discharging capacitor, where $U = U_0 \exp(-t/RC)$.

Such an exponential law is transformed into a linear law on adopting semi-logarithmic coordinates, which lead to

$$\ln y = -\alpha x + \ln y_0.$$

When this rule is represented graphically, one can determine the constant α from the slope $-\alpha$, and the value of y_0 from the intercept $\ln y_0$ on the vertical axis. Both determinations are easy if one uses semi-logarithmic graph-paper with the logarithmic scale along the vertical.

Accordingly, the problem of smoothing the data reduces to determining the regression line, provided one has first subjected them to an appropriate transformation. Admittedly, least-squares fitting to a relation linearized by such a transformation will give results less precise than fitting to the original relation $y = \phi(x)$ itself, but the results will still be pretty good.

We take the discharging capacitor as a concrete example for assessing the method.

The apparatus and the results are sketched in Fig. 3.6. In order to measure the capacitance of the condenser, we first charge it to $U_0 = 100$ V, and then discharge it through the high resistance $R = 1$ MΩ, while recording the variation with time of the voltage $U(t)$ across the capacitor. The results appear in the following table:

t (s)	0	1	2	3	4	5	6	7	8	9	10
U (V)	100	75	55	40	30	20	15	10	10	5	5

Our task is to determine the capacitance C.

The rule for the discharge is $U = U_0 \exp(-t/RC)$, and we shall write it as $U = U_0 \exp(-\alpha t)$. It follows that

$$\ln U = -\alpha t + \ln U_0.$$

We aim for the coefficient $\alpha = 1/RC$, and take the occasion to present both of the simple methods one might use, one graphic and the other algebraic.

The *graphical method* is easy, given semi-logarithmic graph-paper. On this we indicate time horizontally (from 0 to 10 s), and voltage vertically (from 1 to 100 V on a logarithmic scale), as in Fig. 3.7. Then we draw the best straight line we can through all the points,

$$\ln U = -\alpha t + \ln U_0,$$

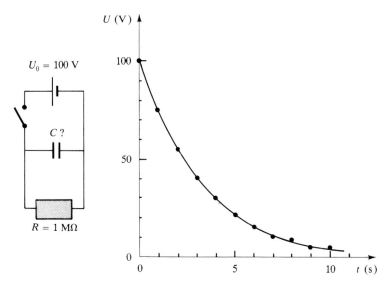

Fig. 3.6 Discharge of a capacitor.

judging simply by eye. Notice that $t = RC$ entails $\alpha t = 1$ and $U/U = e^{-1} = 1/e$. It follows that for $t = RC$ one has $U = 100/e \approx 36.79$. Directly from the graph one estimates that $U = 36.79$ corresponds to $t = 3.2$ s. Hence $RC = 3.2$; in view of $R = 10^6 \ \Omega$, this entails $C = 3.2 \times 10^{-6}$ F, or in other words $C = 3.2 \ \mu F$.

Note in passing that $RC = 3.2$ yields $\alpha = 1/RC = 0.3125$.

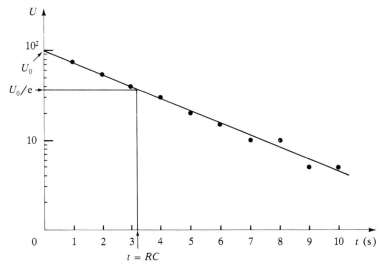

Fig. 3.7 Using a semi-logarithmic scale.

Not all the digits quoted for α are significant, but they will serve in comparing the graphical with the algebraic method.

The *algebraic method* consists in the transformation followed by linear regression. In principle it entails more calculation, because it is somewhat more sophisticated; but in fact an HP-32E calculator can deal with it in short order. The internal programming of the calculator is such that the transformation is implemented simply by pressing the key LN (natural logarithm), and linear regression by the key LR. Appendix 4 gives the sequence of key-strokes and the corresponding readings. At the end of this miniprogram one has three results:

- The vertical intercept $\ln U_0 = 4.6130$; this entails $U_0 = 100.79$ V, which is close enough to the experimental value $U_0 = 100$ V to serve as a check on the calculation.

- The requisite value of the coefficient $-\alpha = -0.3126$, which entails $RC = 1/\alpha = 3.1990$, and thence $C = 3.2 \ \mu F$.

- The correlation coefficient $\rho = -0.9950$. This is very close to -1, reassurance that the appeal to linear regression was quite proper.

It remains only to compare the two methods.

First, we must admit that we started with the graphical method in order that its output should not be influenced by the algebraic result.

We have found

graphical method:	$\alpha = 0.3125$;
algebraic method:	$\alpha = 0.3126$.

Though the algebraic method does represent a refinement, we conclude that the graphical method is already very good. But it must be stressed that using the calculator is at least as fast as drawing the straight line on graph-paper (if not faster).

One must add that, applied to the law linearized by our transformation, least-squares fitting gives a result slightly different from what it would have given applied directly to the original exponential decay law. A complete least-squares calculation with the original data is perfectly feasible, though somewhat lengthy. It would yield $\alpha = 0.307$, which entails $C = 3.25 \ \mu F$. The two methods above, both of them using linearization, gave $C = 3.2 \ \mu F$. In this experiment, therefore, if we wish to stick to simple methods, then we must accept an additional imprecision of $0.05/3 = 1.7\%$. This is acceptable in a teaching laboratory, but would not be so in metrology; however, experimenters there would already be provided with a program on a dedicated computer, capable of producing rapid least-squares fits to any physical law, however complicated.

Such transformations trace their pedigree from Moseley, who in 1913 discovered the connection, now named after him, between the frequency v of X-rays (in a given series) emitted by a chemical element, and the serial number Z of the element in the periodic table. The exact relation between these magnitudes is far from apparent from a simple plot of one against the other; but it does become evident if as in Fig. 3.8 one plots $v^{1/2}$ against Z, corresponding to the transformation $v \rightarrow v^{1/2}$. Then the law becomes linear, taking the form

$$v^{1/2} = k(Z - \alpha).$$

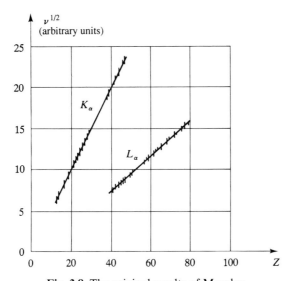

Fig. 3.8 The original results of Moseley.

This very simple relation contains a very great deal of physical information.

1. First, the proportionality of v to Z^2 suggests that the underlying energy is electrostatic and Coulombic, conformably with the atomic model featuring a nucleus having charge $+Ze$ surrounded by Z electrons each having charge $-e$. From this viewpoint Z becomes more than just the serial number of the element in this particular classification; nowadays one calls it the atomic number in order to stress that it actually describes the structure of an atom of the element in question.

2. Next, the need to replace the actual number Z by the effective number $Z - \alpha$ suggests the operation of some screening effect, represented by the screening constant α which is just the intercept on the Z-axis. For the K series

its value is 1, which is small because the K electrons are very close to the nucleus. For the L series by contrast α equals 7.4, because the K electrons interpose between the nucleus and the L electrons.

3. Finally, the transformed relation allows one to clarify the position of the lanthanides, and to attribute correct Z values to the heavy elements.

For so elementary a transformation this is pretty good!

3.7 Conclusions

Techniques of measurement improve spectacularly with the passage of time. This is well illustrated by the evolution of the values found for the speed of light in vacuo.

- The first measurement (1676) gave $c = 214\,000\,000$ m s^{-1}, with a precision† of 3.5.

- A recent measurement (1972)‡ gave $c = 299\,792\,458 \pm 1.2$ m s^{-1}, with a precision of 2.5×10^8.

Thus one can hardly be surprised at the adage popular in laboratories, asserting that 'in the experimental study of any phenomenon, the role of measurements undergoes an evolutionary change: in the first year they are tools of discovery, in the second year they serve as standards, and in the third year they are used to study the background noise.'

Though marginally over-optimistic as regards the time scale, the adage does very vividly reflect the progress of measurement in physics.

Appendices to Chapter 3

1 Estimate of mean and of standard deviation

We have used an HP-32E calculator to determine the mean and the standard deviation of the set of 15 length measurements quoted in Section 3.5. The sequence of key-strokes and the calculator displays are given in the following table.

---

† Precision here is defined by $c/\Delta c$.

‡ Since the most recent meeting of the International Conference on Weights and Measures, in 1983, one no longer measures c, which has now become the natural standard for defining the metre from the second, through the relation $l = ct$.

Key-stroke	Display	Key-stroke	Display
f CLEAR	0.0000	1628 Σ+	9.0000
1627 Σ+	1.0000	1628 Σ+	10.0000
1627 Σ+	2.0000	1629 Σ+	11.0000
1627 Σ+	3.0000	1629 Σ+	12.0000
1628 Σ+	4.0000	1629 Σ+	13.0000
1628 Σ+	5.0000	1629 Σ+	14.0000
1628 Σ+	6.0000	1631 Σ+	15.0000
1628 Σ+	7.0000	f $\bar{x}$	1628.2667
1628 Σ+	8.0000	g s	1.0328

2 Student's distribution for $n-1$ degrees of freedom

Student's table is shown below, followed by instructions for applying it to a set of n measurements of the variable x.

n	t_β $\beta=0.95$	t_β $\beta=0.99$	n	t_β $\beta=0.95$	t_β $\beta=0.99$
2	12.7	63.7	12	2.20	3.11
3	4.30	9.93	14	2.16	3.01
4	3.18	5.84	16	2.13	2.95
5	2.78	4.60	18	2.11	2.90
6	2.57	4.03	20	2.09	2.86
7	2.45	3.71	30	2.04	2.76
8	2.37	3.50	50	2.01	2.68
9	2.31	3.36	100	1.98	2.63
10	2.26	3.25	∞	1.96	2.58

Number of measurements: n

Estimate of $\bar{x}$: $\bar{x}^* = (1/n) \sum_i x_i$.

Estimate of σ: $\sigma^* = \left\{ \dfrac{[\sum_i (x_i - \bar{x}^*)^2]}{(n-1)} \right\}^{1/2}$.

Confidence level: β.

Confidence interval: $\pm t_\beta \sigma^* / n^{1/2}$.

Display format: $\bar{x} = \bar{x}^* \pm t_\beta \sigma^* / n^{1/2}$,
at confidence level β.

Eventually this result must be corrected for systematic deviation, by appeal to prior calibration.

3 Determination of the regression line

We have determined the linear regression for the 12 experimental points quoted in Section 3.4, using an HP-32E calculator. The sequence of key-strokes and the calculator displays are shown in the following table.

Key-stroke	Display	Key-stroke	Display
f CLEAR	0.000000	0.10 $\Sigma +$	7.000000
4.49 ENTER	4.490000	4.27 ENTER	4.270000
0.01 $\Sigma +$	1.000000	0.13 $\Sigma +$	8.000000
4.47 ENTER	4.470000	4.22 ENTER	4.220000
0.02 $\Sigma +$	2.000000	0.16 $\Sigma +$	9.000000
4.45 ENTER	4.450000	4.15 ENTER	4.150000
0.03 $\Sigma +$	3.000000	0.20 $\Sigma +$	10.000000
4.43 ENTER	4.430000	4.06 ENTER	4.060000
0.04 $\Sigma +$	4.000000	0.25 $\Sigma +$	11.000000
4.39 ENTER	4.390000	3.97 ENTER	3.970000
0.06 $\Sigma +$	5.000000	0.30 $\Sigma +$	12.000000
4.36 ENTER	4.360000	f LR	4.502392
0.08 $\Sigma +$	6.000000	$x \rightleftarrows y$	-1.774421
4.32 ENTER	4.320000	g I	-0.999818

4 Transformation followed by linear regression

The following table shows the sequence of key-strokes, on an HP-32E calculator, appropriate to the data quoted in Section 3.5.

Key-stroke	Display	Key-stroke	Display
f CLEAR	0.0000	15 f LN ENTER	2.7081
100 f LN ENTER	4.6052	6 $\Sigma +$	7.0000
0 $\Sigma +$	1.0000	10 f LN ENTER	2.3026
75 f LN ENTER	4.3175	7 $\Sigma +$	8.0000
1 $\Sigma +$	2.0000	10 f LN ENTER	2.3026
55 f LN ENTER	4.0073	8 $\Sigma +$	9.0000
2 $\Sigma +$	3.0000	5 f LN ENTER	1.6094
40 f LN ENTER	3.6889	9 $\Sigma +$	10.0000
3 $\Sigma +$	4.0000	5 f LN ENTER	1.6094
30 f LN ENTER	3.4012	10 $\Sigma +$	11.0000
4 $\Sigma +$	5.0000	f LR	4.6130
20 f LN ENTER	2.9957	$x \rightleftarrows y$	-0.3126
5 $\Sigma +$	6.0000	g r	-0.9950

Further reading

See Moseley (1913, 1914) in the Bibliography.

James Clerk Maxwell. Scottish physicist; born in Edinburgh, died in Cambridge (1831–1879). A brilliant theorist, attracted by deep generalizations, he writes down in 1864 the fundamental equations of classical electromagnetism, which still bear his name. He is as important to the history of physics as Newton is in the seventeenth century and Einstein in the twentieth. With his kinetic theory of gases in 1859, he is the first to progress significantly beyond the theory of games, through his quantitative use of chance in a theory of physical phenomena. (Palais de la Découverte)

4

Maxwell, or probabilities as a matter of ignorance

Est-il des petits corps
plus grossier assemblage?†
Molière

The concept of atoms as elementary grains of matter stems from the Greeks, and more particularly from the teachings of Democritus, twenty-five centuries ago. But of the reasons adduced by its originator we know nothing. At best, we can speculate that he was thinking about the problem of the interpenetrability of two fluids: on mixing milk and wine one obtains an effectively homogeneous pink liquid, and the only easy way to understand this is to picture both the milk and the wine as consisting of elementary particles, white and red respectively. It is these indivisible elementary particles that the Greeks called atoms.

But the true incorporation of atomic theory into science dates from 1806, being due to the English chemist John Dalton. For him, the atomic hypothesis is a model from which to predict the mass-combination rules for the chemical elements. The atoms are elementary particles characterizing the elements, and combine to form molecules, which in turn characterize chemical compounds. This model was to be taken up by the Scottish physicist James Clerk Maxwell when in 1859 he created the kinetic theory of gases. His proposal of a very simple yet very effective model for a perfect gas was a decisive step in the long progression, from Democritus to Jean Perrin, which eventually persuaded scientists to accept the existence of atoms as a demonstrated fact of experience.

4.1 The model for a perfect gas

This model is predicated on three assumptions which then serve to frame three definitions.

- *Structure*: the molecules are hard spheres, and the mean distance between them is much greater than their diameters.

---

† Did small things ever form a ruder crowd?

- *Interactions*: molecular collisions are elastic; between two successive collisions every molecule moves without interacting, that is at constant speed in a straight line.
- *Initial conditions*: the positions and velocities of the molecules are distributed at random.

We have chosen to start by describing the model, in order to stress its simplicity; but it is of course its validity that will concern us most. This, as for all physical models, is not something to be asserted from the outset; it is, rather, endorsed progressively in the light of experimental facts and of the physical insights of the theorist.

● That gases consist of molecules is suggested by studies of their compressibilities. At room temperatures the best-known gases (oxygen, nitrogen, etc.) obey Boyle's law (also called the Boyle–Mariotte law), which for a given mass of gas prescribes the relation

$$PV = \text{constant}.$$

This relation suggests that the structure of all such gases is the same. By contrast, if a gas is cooled sufficiently, it turns into a liquid, which is far more dense and practically incompressible. It is these observations that lead to the model: the molecules are hard spheres far apart, which explains why gases are so compressible. In the liquid on the other hand the hard spheres are in contact, which explains the incompressibility. As to the spherical shape, it is a natural choice in view of its symmetry and in the absence of any other information; we remain free to refine the model later by choosing more elaborate shapes.

● The interactions are governed by classical mechanics. Between collisions every molecule moves at constant speed in a straight line, in accordance with Newton's first law. As to the collisions, they are elastic, because the spheres are hard, or in other words impenetrable and rigid. Our experience with billiard balls illustrates these ideas very clearly (see Appendix 1).

We have now sifted the physics of the first two assumptions underlying the model; both seem perfectly natural when viewed from the state of the subject in 1859. The third assumption by contrast is a remarkable conceptual leap, on which we must now focus our attention.

A priori, the choice of initial conditions presents a very considerable problem. To see this, envisage one mole of a gas, that is a collection of 6×10^{23} molecules in a rigid container, and examine the possibility of representing their behaviour through some deterministic theory.

1. We would need to know the initial position and initial velocity of every

molecule, that is six data (x, y, z, v_x, v_y, v_z) for each. For one mole of gas this makes 3.6×10^{24} initial data altogether.

2. To perform the calculation we should need to solve a system of 1.8×10^{24} equations, featuring these 3.6×10^{24} initial conditions. The result would consist of 3.6×10^{24} numbers, representing the state of the system at some given later time t.

3. Finally, we should need to calculate the readily measurable macroscopic variables, like the pressure, as appropriate sums over these 3.6×10^{24} output values.

Given our knowledge of twentieth-century physics, we can see that attempts to implement such a deterministic programme must run into several insurmountable obstacles. First, there is the physical obstacle represented by the Heisenberg inequalities, that is by the impossibility of measuring simultaneously the position and the velocity of a molecule (see Chapter 7). The most we can assert is that for a perfect gas the distances between molecules are much greater than their diameters, which brings us under the provisions of the classical approximation to quantum mechanics. If we wish to introduce the simultaneous position and velocity of a molecule into a classically deterministic calculation, then we must bear in mind that this can be done only as an approximation. Next, there is a technical obstacle, because such calculations would be so lengthy as to be totally impracticable even on the fastest present-day computers. One can get some idea of this by noting that just in order to store the data one would need a magnetic tape ten thousand light-years long; such a tape when extended would stretch to the limits of our galaxy, and would occupy quite a volume even when spooled. Finally, there is a methodological objection which one can formulate as follows: since our aim is to calculate macroscopically measurable quantities from microscopic data, might one not do this more easily than through such a deterministic theory?

Though Maxwell in 1859 was not fully aware of all these unavoidable obstacles and objections, he could not but realize that in practice such a calculation is impossible. It was in this context that he made the conceptual leap of introducing into the theory the notion of the unpredictable. Since the molecules and their collisions are so numerous, and the velocities so varied, and since our ignorance of the initial conditions is almost total, Maxwell postulated that positions and velocities are distributed at random; and he was confident that this assumption would describe the gas adequately and would allow one to calculate the mean values of the macroscopic variables. His breathtaking intuition was confirmed half a century later by the work of Albert Einstein (1905) and of Jean Perrin (1908) on Brownian motion. Of course, it is not enough to assert merely that the distribution is random; one must also specify the right distribution law, and it is here that one comes to appreciate the full scope of Maxwell's physical insight.

4.2 The probabilistic assumptions introduced by Maxwell

In order to find the requisite distribution we define two reference frames (Fig. 4.1):

- *coordinate space*, fixed with respect to the container, which allows us to specify the position of a molecule by means of a vector r having the components x, y, z;

- *velocity space*, which is more of an abstraction, and which allows us to specify the velocity vector v of a molecule by means of its three components v_x, v_y, v_z.

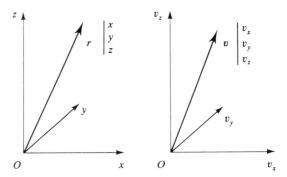

Fig. 4.1 Coordinate space and velocity space.

We are now in a position to state the three probabilistic assumptions proposed by Maxwell.

1. It is natural to assume that *the distribution of the molecules in (coordinate) space is uniform*, because one can see no reason why they should accumulate in any particular region of the container.† Mathematically speaking, this means that the probability density assigned to r is just a constant

---

† In this first approach, we neglect the effects of gravity.

independent of r. Hence it can depend only on v, and we write it as $F(v)$ or as $F(v_x, v_y, v_z)$. In other words, $F(v_x, v_y, v_t) \Delta v_x \Delta v_y \Delta v_z$ is the probability that a molecule chosen at random has a velocity vector pointing into the volume element $\Delta v_x \Delta v_y \Delta v_z$ around v.

2. Intuitively, the fact that there are so many molecular collisions suggests the assumption that *the three components of the velocity are mutually independent*: the collisions shuffle the velocities, so to speak, so effectively that each component becomes totally independent of the other two. This assumption is less compelling than the first, and was much debated by physicists at the time. As so often in physics, its only real justification lies in its success. It entails a great simplification in the mathematical form of the probability density $F (v_x, v_y, v_z)$: the three variables v_x, v_y, v_z are independent, and the three partial distributions take the forms $f_1(v_x)$, $f_2(v_y)$, $f_3(v_z)$, each a function of only one component. This makes it possible to separate the variables, and to write the overall probability density as

$$F (v_x, v_y, v_z) = f_1(v_x), \; f_2(v_y), \; f_3(v_z).$$

3. Symmetry makes it natural to assume that *the probability distribution is isotropic*, since one cannot imagine that any directions are privileged *a priori*. Mathematically speaking this entails that the probability density assigned to v depends only on the magnitude v of the vector v, and that the three partial distributions f_1, f_2, f_3 all have the same form f, because none of the directions plays any special role. Accordingly, the probability density simplifies even further, and reads

$$F(v) = f(v_x)f(v_y)f(v_z).$$

This very elementary condition then suffices to determine the probability $F(v)$ assigned to the vector v (see Appendix 3).

To summarize, Maxwell does not reject determinism as such, but he mitigates our ignorance of the initial conditions by introducing assumptions of a probabilistic nature. This places us squarely within a framework of *probabilities through ignorance*.

4.3 Pressure and temperature

If we now consider a given mass of gas confined inside a box, we can envisage its microscopic structure quite clearly: vast numbers of molecules are rushing around in all directions, subjecting the walls of the container to an intense bombardment in the process. It is the average overall effect of this bombardment that we represent as the pressure exerted by the gas. It is an easily measurable macroscopic variable; our aim is to calculate the pressure in terms of the microscopic properties of the ideal gas as defined in Maxwell's model.

The calculation of the mean pressure is short and simple (see Appendix 2); the result reads

$$\bar{P} = \tfrac{1}{3}\bar{n}m\overline{v^2},$$

where $\bar{P}$ is the mean pressure of the perfect gas, $\bar{n}$ is the mean number of molecules per unit volume, m is the mass of one molecule, and $\overline{v^2}$ is the mean-squared (average value of the squared) molecular velocity.

This leads us quite naturally to introduce the root-mean-square velocity v_{rms} through

$$v_{rms} = (\overline{v^2})^{1/2}.$$

Moreover, as we shall see near the end of this section, the relative fluctuations of the instantaneous pressure are very weak. This leads us to identify the mean pressure $\bar{P}$, which is a calculated quantity, with the experimental pressure P measured with a manometer.

The equation of state of a perfect gas is a macroscopic law derived from experiment; for one mole it reads $PV = RT$. We shall exploit it by writing the pressure as

$$P = RT/V,$$

where P is the pressure of the perfect gas, R is the perfect-gas constant $(8.314\,510\ \mathrm{J\ K^{-1}\ mol^{-1}})$, T is the absolute temperature, and V is the molar volume.

We can now establish a correspondence between microscopic and macroscopic variables. To this end, it suffices to equate the mean pressure $\bar{P}$, from our model of the perfect gas, with the pressure P from the equation of state:

$$\tfrac{1}{3}\bar{n}m\overline{v^2} = RT/V.$$

On noting that $(\overline{v^2})^{1/2} = v_{rms}$, the root-mean-square velocity, $\bar{n}V = N$ (Avogadro's number), and $Nm = M$ (the molar mass), a short calculation yields

$$v_{rms} = (3RT/M)^{1/2},$$

where v_{rms} is the root-mean-square velocity ($\mathrm{m\ s^{-1}}$), R is the perfect-gas constant, T is the absolute temperature (K), and M is the molar mass (kg).

An alternative form reads

$$v_{rms} = (3kT/m)^{1/2}.$$

This form features Boltzmann's constant $k = R/N$, where $k = 1.380\,658 \times 10^{-23}\ \mathrm{J\ K^{-1}}$.

Accordingly, Avogadro's constant N governs the relation between the macroscopic level, characterized by R and M, and the microscopic level, characterized by k and m.

A numerical example helps one to appreciate this relation. For air at 20°C, that is at 293 K, we find

- for nitrogen, $v_{rms} = 510 \text{ m s}^{-1}$;

- for oxygen, $v_{rms} = 480 \text{ m s}^{-1}$.

The physical significance of temperature is now manifest. On the one hand, it is the macroscopic physical variable measured by a thermometer. On the other hand, microscopically it corresponds to the mean-squared speed, which we can evaluate in terms of the two physical constants R and M†. The absolute temperature thus becomes our scale for molecular agitation; that is why one speaks of *thermal agitation* in describing the ceaseless random motions of molecules.

Our picture of a perfect gas on the molecular scale is now quite clear; we have sketched it in Fig. 4.2.

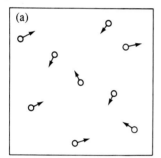 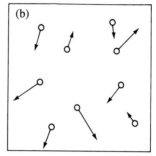

Fig. 4.2 A perfect gas shown on a molecular scale: (a) at low temperature; (b) at high temperature. (The arrows indicate the velocities.)

The fluctuations of the pressure are also calculable, at least as regards their order of magnitude. For example, for a molar volume under normal conditions, we can calculate the number of impacts per second on an area of 10 cm × 10 cm. The interval (1 s) and the area (100 cm²) are typical in pressure measurements. The number of impacts thus calculated is of order 10^{26}. Its fluctuations are well represented by the standard deviation $\sqrt{10^{26}}$ (Section 2.5); they are therefore of order 10^{13}, whence we derive a relative deviation of 10^{-13}. The relative deviation in classic pressure measurements is

---

† The constants R and M are measured through macroscopic phenomena: R through the compressibility and the thermal expansion of gases, and M with the chemical balance.

of the order of 10^{-2} to 10^{-3}, far in excess of the fluctuations in the number of impacts, which it is therefore permissible to neglect. This observation amounts to an *a posteriori* justification of Maxwell's model of the perfect gas. On the one hand, we have the deterministic theory which, if it were operable (as it is not), would yield the instantaneous value of the pressure; on the other hand, we have the readily operable probabilistic theory, which gives the mean value of the pressure; and the relative deviation between these two theories is 10^{-13}. In the first section of this chapter we stressed the need to interrelate only comparable results when substituting chance for determinism. The word 'comparable' has now been quantified: it is indexed by a relative fluctuation of 10^{-13}.

At this stage of the discussion, our model of a perfect gas has satisfied the first condition for the validity of a physical theory: it explains in coherent fashion phenomena already familiar to experiment, namely pressure and temperature, and is not therefore contradicted by the physics we already know. It must still be shown to satisfy the second condition, by making new predictions verifiable through experiment; and that is what we now proceed to do, by determining the distribution law for the molecular velocities.

4.4 The velocity distribution

In order to determine this distribution we need the three probabilistic assumptions already listed: uniform distribution in space, mutual independence of the three velocity components, and isotropy as regards the directions of the velocities. The end-result for the velocity distribution (Appendix 3) reads

$$\rho(v) = 4\pi \left(\frac{m}{2\pi kT} \right)^{3/2} \left[\exp\left(-\frac{mv^2}{2kT} \right) \right] v^2,$$

where $\rho(v)$ is the probability density assigned to the speed v (the magnitude of the velocity), T is the absolute temperature of the perfect gas, m is the mass of a molecule, and k is Boltzmann's constant.

This way of displaying the result calls for an immediate technical comment: $\rho(v)$ is a probability density, which means that, on measuring the speed of a molecule chosen at random, there is a probability $\rho(v) \Delta v$ for this speed to lie between v and $v + \Delta v$.†

Some remarks about the physics are now in order.

---

† Do not confuse the distribution $F(\mathbf{v})$ for the velocity vector $\mathbf{v}$ with the distribution for the speed v (see Appendix 3).

1. The probability density $\rho(v)$ depends on the temperature but not on the pressure. This is true only in so far as the gas may be considered as perfect, that is only while the temperature remains well above the critical temperature.

2. The density $\rho(0)$ is zero, which means that no molecules are at rest. There can, however, be a few molecules whose speeds are very low.

3. The density $\rho(v)$ is very small, though non-zero, when v becomes much larger than $(\overline{v^2})^{1/2}$. For the Earth's atmosphere, for instance, this implies that there are always some molecules with speeds higher than $11\ 300\ \mathrm{m\ s^{-1}}$, which is the escape velocity. Hence there are always some molecules escaping from the atmosphere, which is slowly thinning out in consequence. There is no need for immediate alarm: the lifetime of the atmosphere is measured in thousands of millions of years [24]. Mankind has some time in which to find technical solutions to the problem.

The graph of the velocity distribution is a curve so familiar to physicists that it has come to be called the Maxwellian distribution in honour of its inventor (see the broken curve in Fig. 4.3).

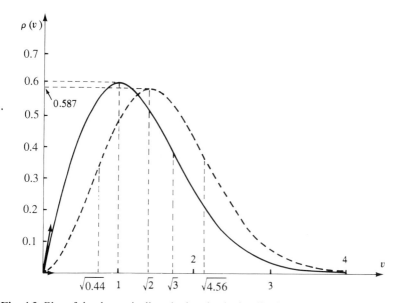

Fig. 4.3 Plot of the theoretically calculated velocity distribution in a two-dimensional gas (continuous curve) and in a three-dimensional gas (broken curve). Velocities are shown horizontally in units of $(kT/m)^{1/2}$; probability densities are shown vertically in units of $(m/kT)^{1/2}$.

Its shape should be noted: horizontal tangents at $v=0$ and $v=(2kT/m)^{1/2}$, and a horizontal asymptote as v tends to infinity. Note also that the maximum corresponds to the most probable velocity $v_{mp}=(2kT/m)^{1/2}$, which differs from the root-mean-square velocity $v_{rms}=(3kT/m)^{1/2}$; this difference reflects the fact that the distribution is asymmetric.

The theory has now spoken and its consequences have been expressed in simple form. It remains to compare prediction with experiment.

4.5 The theory tested by experiment

A great deal of work has been devoted to testing the predicted velocity distribution; here we settle for quoting just one experiment, performed in 1959 by P. M. Marcus and J. H. McFee [17], whose results are particularly precise and convincing. Their layout is sketched in Fig. 4.4.

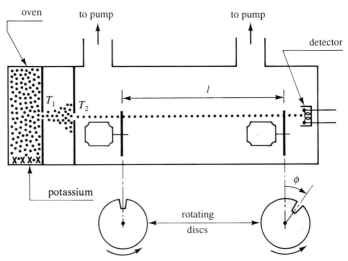

Fig. 4.4 Sketch of the experiment by P. M. Marcus and J. H. McFee to verify the Maxwell distribution, using a beam of potassium atoms.

The apparatus exploits the technique of molecular beams, using potassium. Like all the alkali metals, potassium is monatomic in the gas phase; accordingly, the beam traversing the apparatus consists of neutral atoms. It is produced by an oven which vaporizes the potassium under reduced pressure, at a temperature of 157°C. Atoms of the gas thus produced leave the oven through a hole T_1, and are then channelled through a second hole T_2; rapid secondary pumping establishes dynamical equilibrium at a pressure of 0.84×10^{-3} torr† near the oven, and of 0.5×10^{-7} torr beyond T_2. In this way one obtains a beam of neutral atoms travelling through the chamber in a straight line from the diaphragm T_2 to the detector. The detector is a filament glowing red; on colliding with it the neutral potassium atoms loose their outermost electron to become positive ions. These ions are captured by a neighbouring electrode kept at a negative potential with respect to the filament, so that what one measures is an ionic current corresponding to the flux of neutral atoms. In order to measure the speeds of the atoms in the beam, one uses two discs whose planes are perpendicular to the beam, a distance l apart, each disc being provided with a notch that represents an opening of 4°. The discs are mounted on the axles of two synchronous motors rotating in phase at the same angular velocity ω (8000 rpm); the phase difference between the notches is constant, but adjustable. Each disc shuts off the atomic beam completely, except during the brief interval while the notch is crossing the beam path.‡

Consider now an atom having velocity v, and crossing the notch in disc D_1 at time t_1; then, given the phase lag ϕ, the notch in disc D_2 crosses the beam path at a time t_2 such that

$$\phi = \omega(t_2 - t_1).$$

If the atom gets through the notch in disc D_2, then it must have traversed the intervening distance 1 in precisely the time $t_2 - t_1$; hence its velocity must be given by

$$l = v(t_2 - t_1),$$

or better, by $v = l\omega/\phi$.

Accordingly, the pair of co-rotating discs acts as a *velocity selector* governed by the transit time $t_2 - t_1$, while the detecting filament measures the relative abundance of atoms travelling at the selected speed. By measuring this abundance as a function of the phase lag ϕ, one can study the velocity distribution.

The experimental results are shown in Fig. 4.5. The phase difference between the discs is plotted horizontally, and the beam intensity at the detector vertically, both in arbitrary units. There are many (62) experimental points, and their scatter is evidently slight. They are compared with the theoretical distribution, which for an atomic beam takes the form

---

† 1 torr = 1 mm of mercury $\approx$ 133 Pa.

‡ We have described the principle of the experiment as simply as possible. The actual apparatus is modified in two ways so as to maximize the intensity of the atomic beam reaching the detector. First, the holes T_1 and T_2 are replaced by slits. Second, each disc carries six 4° notches, spaced at 60°.

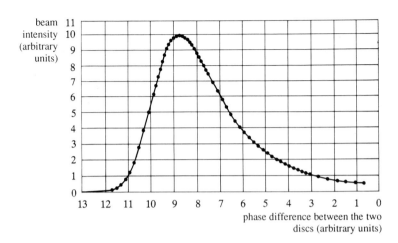

Fig. 4.5 Experimental verification of the Maxwell velocity distribution, using a beam of potassium atoms emitted at temperature of 157°C (after P. M. Marcus and J. H. McFee).

$$\rho'(v) = av\rho(v) = av^3 \exp(-bv^2).$$

The agreement between theory and experiment is quite remarkable and well worth stressing.

4.6 Simulating a perfect gas

As regards the researcher the cycle is complete, since we have gone through all the stages of the scientific process:

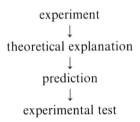

The teacher, however, still has a serious problem, in that the atoms adduced by our model are invisible, and a velocity distribution in particular is something of an abstraction. All this would become much more telling if we could construct a macroscopic system governed by the same laws as the microscopic model; we could then exhibit on a laboratory scale a simulation of what actually happens microscopically. The first idea that comes to mind is to use a container with very rigid walls, supply it with ball-bearings made of very hard steel, pump out the air, and put the whole contraption on board a space lab in order to eliminate the effects of gravity. It is clear that the pursuit of such a programme might be awkward, and it seems preferable to look for some sensible compromise, likely to prove more readily realizable.

It turns out that an air table fits the bill, provided we are willing to make one small concession: the gas will be simulated in two dimensions instead of three. This change in the geometry changes the velocity distribution, which now reads (see Appendix 3)

$$\rho(v) = \frac{m}{kT} \left[\exp\left(-\frac{mv^2}{2kT} \right) \right] v.$$

The main physical characteristics of the model, however, are maintained. The molecules are mimicked by hard discs; floating on an air cushion, these discs move without friction, in straight lines at constant speeds; and collisions between them are elastic. We shall assess the quality of the simulation at the end of this section, and proceed to present the experimental data.

The air table we use is flat and horizontal, and forms a square with edge length 1.20 m. It has very many small performations, 15 mm apart on average, which serve as exit jets for air forced under the table by a compressor. On this table we place a (movable) rectangle, 0.75 m × 0.53 m, which represents the walls of the container. Piano wires stretched along its edges ensure that impacts are highly elastic. The rectangle rests on two large wings which allow it to float smoothly on the air cushion; it is connected through a driver arm to a small electric motor which can cause it to oscillate with variable period. The mobile discs are 4 cm in diameter and 1 cm thick, all made of the same hard plastic material (nylon or teflon). The discs are well supported, because even with diameters this small, there are still on average seven performations under each disc. The process is recorded using a Polaroid camera and an electronic stroboscope.

The experimental procedure is very simple. The table is carefully levelled on screw feet, and we put on it 25 'monatomic molecules', that is 25 discs. Next we establish a regime of 'thermal agitation' (air cushion and vibrating walls), and install the camera (looking vertically downward) and the stroboscope (oblique angle of incidence). We open the camera shutter in the dark, and illuminate the (thermally agitated) table with a sequence of five flashes, one-eighth of a second apart. The resulting photograph allows each trajectory to be identified on sight. By locating the initial and final positions of the disc centres (first and fifth (i.e. last) flashes respectively), we can determine the distances travelled in half a second (allowing for the enlargement), whence we can easily determine the speeds. Note that the velocities will be randomly distributed in direction. We have analysed 32 exposures, measuring 461 velocities in the process. The results are given in the table below.

Number of the interval (widths $\Delta v = 1$ cm s^{-1})	0	1	2	3	4	5	6	7	8	9	10
Number of measured velocities (Δn)	14	6	26	43	54	48	55	40	29	35	31

Number of the interval (widths $\Delta v = 1$ cm s^{-1})	11	12	13	14	15	16	17	18	19	20	21
Number of measured velocities (Δn)	20	17	12	12	6	6	4	0	1	1	1

The observed distribution is shown in Fig. 4.6. The velocity in centimetres per second is plotted horizontally; vertically we plot the number Δn of results falling into each interval of $\Delta v = 1$ cm s^{-1}. Each point also shows the standard deviation $\pm (\Delta n)^{1/2}$ (see Section 3.3).

The corresponding theoretical curve takes the form (see Appendix 4)

$$\Delta n = 1.65 \frac{(\Delta n)_0}{v_0} \left[\exp\left(-\frac{v^2}{2v_0^2} \right) \right] v \, \Delta v.$$

It depends on two parameters, v_0 and $(\Delta n)_0$, which feature as the coordinates of the maximum. In principle, the parameters should be determined through a least-squares fit to the data; but unfortunately this is very time-consuming except with a computer. Hence we indicate a simple fitting procedure in common use.

A first approximation is obtained from the centre of gravity of the three highest points, and yields $v_0 = 5.5$ and $(\Delta n)_0 = 52$. Next, one evaluates and plots, point by point, a first 'theoretical' curve, which gives good agreement near the maximum, but which is slightly shifted at high velocities. The position of the maximum is now readjusted so as to diminish this shift, say by choosing $v_0 = 5.8$ and $(\Delta n)_0 = 51$. The eventual theoretical curve is the solid line in Fig. 4.6. It tallies with the experimental points very well: remember that, given the standard deviation, one experimental point in three should, on average, lie outside the curve.

Is there any way to diminish the statistical deviations appreciably, so as to get an even more searching test of the theory? It requires on average four hours of analysis to secure the data given above. In order to diminish the statistical

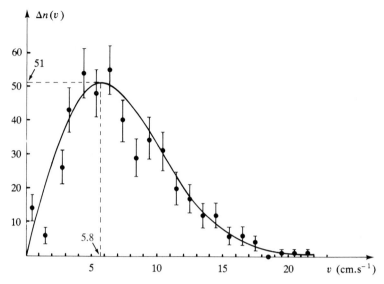

Fig. 4.6 Experimental results for the velocity distribution in a 'perfect two-dimensional gas'. The standard deviation $\pm(\Delta n)^{1/2}$ is indicated for each point. The theoretical distribution is calculated point by point, starting from an adjustment of the maximum to $v_0 = 5.8$ cm s^{-1} and $(\Delta n)_0 = 51$. It is plotted as the continuous curve.

deviations by a factor of 2, we should need four times the data, and this would take 16 hours. Settling for four hours seems reasonable, since it does yield an excellent result.

Just how valid the simulation is remains to be discussed. We know it is not perfect, because there is still some friction (in spite of the air cushion), and because the collisions are not perfectly elastic (even though the discs are very hard). Hence our set of 25 discs does dissipate energy in the form of heat, and it is a function of the vibrating boundary to reinject mechanical energy into the system so as to maintain it in a state of statistical equilibrium. The simulation of a two-dimensional perfect gas on an air table is not, therefore, compelling experimental proof that the assumptions underlying the model lead to the two-dimensional Maxwell distribution for such a gas. On the other hand, we can reverse the argument, and claim that, if the observed distribution is Maxwellian, then it is so because the macroscopic simulation is reasonably true to the microscopic system, that is to a collection of molecules in thermal agitation. From this point of view, the purpose of the simulation is essentially pedagogical: when learning physics, one always tries, consciously or not, to picture processes on the molecular scale through analogies on the laboratory scale. The simulation we have described has no purpose other than this, which, however, it serves very well.

4.7 Assessment of the model

The purpose of our model for a perfect gas was to take the atomic and molecular theory of the chemists and to exploit it in physics. The success of this programme reinforces the theory by extending its domain of application.

In order to proceed from ideal to real gases, the theory needs to be refined. First, one must allow for the fact that not all gas molecules are monatomic; they can be diatomic, triatomic, or even more complex. Second, one must take into account two factors that cease to be negligible when the pressure is high enough, and the temperature close enough to the critical temperature: namely molecular congestion and intermolecular attraction. They modify the equation of state, and to a good approximation replace the perfect-gas law by Van der Waals' equation for real gases:

$$\left(P + \frac{a}{V^2} \right)(V - b) = RT.$$

This equation, written here for one mole, features two adjustable positive parameters, a and b.

Chance has thus been introduced to good effect; it has allowed us to proceed beyond microscopic determinism, which though unchallenged in principle proves useless in practice. Through our model for a perfect gas, chance acquires its initial status, that of chance as a matter of ignorance.

Appendices to Chapter 4

1 Molecular interactions in the model for a perfect gas

We have modelled the molecules of a perfect gas as hard spheres. This model serves automatically to specify also the interactions between molecules: all collisions are elastic, whether between molecule and molecule or between molecule and wall. It follows that the total kinetic energy that the molecules possess by virtue of their thermal agitation is constant, which is effectively what one observes, in that the gas remains at constant temperature.

In fact the assumption is rather too restrictive as compared with the true state of affairs, because collisions can be either elastic or inelastic, and it is only the average of the total kinetic energy that remains constant. We could, accordingly, adopt this latter and more realistic assumption, without in any way affecting the derivation of the velocity-distribution law. Why then do we prefer to maintain the restriction to purely elastic interactions? There are two reasons, one physical, the other pedagogic.

1. Our minds are naturally receptive to the model of microscopic billiard balls representing atoms interacting elastically; in other words this model appears to be the simplest. It is a general rule of theoretical physics to choose the simpler of two models both yielding the same predictions (in this case, the model with elastic-only collisions in preference to the model with inelastic collisions as well); the customary form of words is that everything happens 'as if'.

2. The model with hard macroscopic spheres is very closely mimicked by hard macroscopic discs in agitation on an air table: and the use of analogies whenever possible is one of the keys to good teaching.

2 Calculation of the mean pressure

Consider a box containing a perfect gas, with a mean number $\bar{n}$ of molecules per unit volume; we aim to calculate the mean force $\bar{F}$ exerted by the gas on an element of the wall having area S. We start by analysing the case where there is just one molecule.

The interaction between the wall and the molecule under study is elastic. Before the collision, the horizontal component of the velocity v is v_x (Fig. 4.7a); after the collision, it is $-v_x$. The force exerted by the wall on the molecule, at any given time during the collision, is

$$-\Delta F = m \frac{dv_x}{dt}.$$

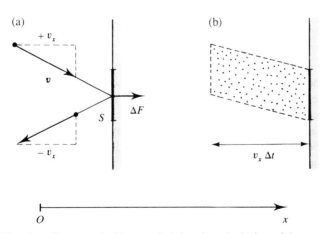

Fig. 4.7 Sketch to illustrate the ideas underlying the calculation of the mean pressure: (a) elastic collision of a molecule with the wall; (b) summation over collisions with velocity v in the time interval Δt.

Hence the force exerted by the molecule on the wall at the same time is

$$\Delta F = -m \frac{dv_x}{dt}.$$

In fact we require not ΔF but its mean value $\overline{\Delta F}$, that is its average over the duration Δt of the pressure measurement. This duration is of the order of 1 s. Hence we write

$$\overline{\Delta F} = -\frac{1}{\Delta t} \int_t^{t+\Delta t} m \frac{dv_x}{dt}\, dt.$$

The integrand is zero over most of the time-interval Δt, except very briefly while the elastic collision is taking place. The integral is evaluated by changing the variable:

$$\overline{\Delta F} = -\frac{1}{\Delta t} \int_{v_x}^{-v_x} m\, dv_x = -\frac{1}{\Delta t}\left[(-mv_x)-(mv_x)\right] = \frac{2mv_x}{\Delta t}.$$

Those molecules that are liable to make a direct impact on the wall have velocities whose horizontal component we write as v_x, in view of our sign convention. There are on average a number $\frac{1}{2}\bar{n}$ of such molecules per unit volume, by virtue of the assumption that velocities are distributed isotropically. Out of this set of molecules, we start by contemplating those having velocity v. Their mean number is $\Delta(\frac{1}{2}\bar{n})$. Which (and how many) of them reach the surface element S during the time interval Δt? All those that are contained in an inclined cylinder of cross-sectional area S and of height $v_x \Delta t$ (Fig. 4.7b). The volume of this cylinder is $Sv_x \Delta t$, and it contains a mean number $\Delta(\frac{1}{2}\bar{n})Sv_x\Delta t$ of molecules. Every one of these molecules exerts a mean force $\Delta\bar{F}$; jointly therefore they exert a force

$$\frac{2mv_x}{\Delta t}\Delta(\tfrac{1}{2}\bar{n})Sv_x\,\Delta t = 2mS\,\Delta(\tfrac{1}{2}\bar{n})v_x^2.$$

Next we must sum over all velocities v that have a positive horizontal component v_x; in other words we must proceed from the set $\Delta(\frac{1}{2}\bar{n})$ to the set $\frac{1}{2}\bar{n}$. Accordingly, the requisite mean value $\bar{F}$ reads

$$\bar{F} = \sum 2mS\,\Delta(\tfrac{1}{2}\bar{n})v_x^2.$$

We rearrange this slightly and obtain

$$\bar{F} = 2mS(\tfrac{1}{2}\bar{n})\sum \frac{\Delta(\frac{1}{2}\bar{n})}{\frac{1}{2}\bar{n}}\, v_x^2 = \bar{n}mS\sum \frac{\Delta(\frac{1}{2}\bar{n})}{\frac{1}{2}\bar{n}}\, v_x^2.$$

The sum $\sum$ is simply the definition of the average value of v_x^2, which we write as $\overline{v_x^2}$; here it is the average over half the molecules, namely over those that have positive v_x. However, this restriction makes no difference to the result, because the velocities $\boldsymbol{v}$ are distributed isotropically, which entails in particular that the distribution of the v_x is symmetric around the value $v_x = 0$. Thus we may write

$$\bar{F} = \bar{n} m \overline{v_x^2} S.$$

The mean pressure follows immediately through the relation $\bar{P} = \bar{F}/S$:

$$\bar{P} = \bar{n} m \overline{v_x^2}.$$

The isotropy of the velocity distribution leads to a more illuminating expression. From

$$\boldsymbol{v} = \boldsymbol{v}_x + \boldsymbol{v}_y + \boldsymbol{v}_z,$$

we find the squared magnitude of the vector $\boldsymbol{v}$:

$$v^2 = v_x^2 + v_y^2 + v_z^2.$$

Hence the mean values are related by

$$\overline{v^2} = \overline{v_x^2} + \overline{v_y^2} + \overline{v_z^2}.$$

Finally, by virtue of isotropy, we have

$$\overline{v_x^2} = \overline{v_y^2} = \overline{v_z^2},$$

whence

$$\overline{v_x^2} = \tfrac{1}{3}\overline{v^2}.$$

Therefore the end-result for the mean pressure reads

$$\bar{P} = \tfrac{1}{3}\bar{n} m \overline{v^2}.$$

It is important to notice that this derivation, by using finite numbers $\Delta\bar{n}$ and a discrete sum $\sum$, has assumed tacitly that the velocities are quantized, which is indeed the case for a particle confined to a box. Essentially the same derivation could be formulated more classically in terms of differentials dn and a continuous sum $\int$, but the details would be somewhat more troublesome.

3 The Maxwell velocity distribution

We present a somewhat old-fashioned derivation, which does however have the advantage of staying close to the physical assumptions underlying the model. It uses the method of Lagrange multipliers. There exists an alternative derivation, using Gibbs ensembles, more powerful but also more formal, which is why we pass it by [14].

We aim to find the probability density $F(\boldsymbol{v})$, such that

$$F(\boldsymbol{v}) = F(v_x, v_y, v_z) = f(v_x)f(v_y)f(v_z).$$

Given the value v of the speed (i.e. the magnitude of the velocity vector $\mathbf{v}$), isotropy in velocity space implies that both the following conditions are satisfied simultaneously:

$$v_x^2 + v_y^2 + v_z^2 = \text{constant}, \tag{1}$$

$$F(v_x, v_y, v_z) = \text{constant}. \tag{2}$$

Differentiation yields

$$v_x \, dv_x + v_y \, dv_y + v_z \, dv_z = 0, \tag{1'}$$

$$\frac{\partial F}{\partial v_x} \, dv_x + \frac{\partial F}{\partial v_y} \, dv_y + \frac{\partial F}{\partial v_z} \, dv_z = 0. \tag{2'}$$

On the other hand, the mutual independence of the three components v_x, v_y, v_z allows us to write

$$F(v_x, v_y, v_z) = f(v_x)f(v_y)f(v_z),$$

$$\ln F(v_x, v_y, v_z) = \ln f(v_x) + \ln f(v_y) + \ln f(v_z),$$

$$\frac{1}{F}\frac{\partial F}{\partial v_x} = \frac{1}{f(v_x)}\frac{df(v_x)}{dv_x},$$

$$\frac{\partial F}{\partial v_x} = \frac{F}{f(v_x)}\frac{df(v_x)}{dv_x}.$$

The last relation leads to a new form of (2′), and we are left with

$$\begin{cases} v_x \, dv_x + v_y \, dv_y + v_z \, dv_z = 0, & (1'') \\ \dfrac{1}{f(v_x)}\dfrac{df(v_x)}{dv_x} \, dv_x + \dfrac{1}{f(v_y)}\dfrac{df(v_y)}{dv_y} \, dv_y + \dfrac{1}{f(v_z)}\dfrac{df(v_z)}{dv_z} \, dv_z = 0. & (2'') \end{cases}$$

Let us now introduce a Lagrange multiplier, that is a function λ of the three variables v_x, v_y, v_z; it enables us to amalgamate the two equations (1″) and (2″) into

$$\left(\lambda v_x + \frac{1}{f(v_x)}\frac{df(v_x)}{dv_x}\right) dv_x + \left(\lambda v_y + \frac{1}{f(v_y)}\frac{df(v_y)}{dv_y}\right) dv_y + \left(\lambda v_z + \frac{1}{f(v_z)}\frac{df(v_z)}{dv_z}\right) dv_z$$
$$= 0.$$

This can be an identity only if the coefficients of all three differentials dv_x, dv_y, dv_z vanish, whence we have

$$\frac{1}{f(v_x)}\frac{df(v_x)}{dv_x} = \frac{d}{dv_x}\ln f(v_x) = -\lambda v_x,$$

plus the corresponding relations for v_y and for v_z.

The first equation shows that the function λ can depend only on v_x, but must be independent of v_y and v_z. Similar reasoning from the two other equations shows that λ cannot in fact depend on *any* of the three variables v_x, v_y, v_z. Hence λ is a constant, which makes it easy to integrate all three equations. We find

$$\ln f(v_x) = \tfrac{1}{2}\lambda v_x^2 + \ln A,$$

whence

$$f(v_x) = A \exp(-\tfrac{1}{2}\lambda v_x^2),$$

with similar expressions for $f(v_y)$ and $f(v_z)$. Here, A and λ are constants still to be determined.†

As regards the overall probability density $F(v_x, v_y, v_z)$, it is the product of the three partial densities $f(v_x), f(v_y)$, and $f(v_z)$; hence

$$F(v_x, v_y, v_z) = A^3 \exp[-\tfrac{1}{2}\lambda(v_x^2 + v_y^2 + v_z^2)].$$

The derivation is now almost complete, in that the exponential character of the distribution has been established. The only object of the calculations that follow (and which are as long as what has gone before) is to satisfy the boundary conditions, and to verify that microscopic and macroscopic aspects are compatible.

The sign of λ is governed by the norming condition: the integral of the probability density over (all of) velocity space must be 1. The integral reads

$$\iiint A^3 \exp[-\tfrac{1}{2}\lambda(v_x^2 + v_y^2 + v_z^2)] \, dv_x \, dv_y \, dv_z = 1.$$

It converges only if the argument of the exponential is negative, that is only if λ is positive. Accordingly we write $\tfrac{1}{2}\lambda = \mu^2$, whence

$$F(v_x, v_y, v_z) = A^3 \exp[-\mu^2(v_x^2 + x_y^2 + v_z^2)].$$

The constants A and μ are determined by ensuring that, in one mole of a perfect gas, the number of molecules is equal to Avogadro's number N, and that the internal energy is $\tfrac{3}{2}RT$.

• The number of molecules enters through the definition of the probability density:

$$\frac{\Delta N}{N} = A^3 \exp[-\mu^2(v_x^2 + v_y^2 + v_z^2)] \, \Delta v_x \, \Delta v_y \, \Delta v_z,$$

where ΔN is the number of molecules in the volume element $\Delta v_x \, \Delta v_y \, \Delta v_z$ of velocity space.

---

† By isotropy, the constant A is the same for all three variables.

In order to count all the molecules having speeds between v and $v + \Delta v$, we consider the volume element contained between the spheres of radii v and $v + \Delta v$. One finds

$$\frac{\Delta N}{N} = A^3 \exp(-\mu^2 v^2) 4\pi v^2 \, \Delta v,$$

whence

$$\Delta N = 4\pi N A^3 v^2 \exp(-\mu^2 v^2) \, \Delta v.$$

Integration over v from 0 to ∞ yields

$$N = 4\pi N A^3 \int_0^{+\infty} v^2 \exp(-\mu^2 v^2) \, dv.$$

We are faced here with a classic so-called improper integral. Its value is $(1/4\mu^2)(\pi/\mu^2)^{1/2}$, which yields one relation between A and μ:

$$A\pi^{1/2}/\mu = 1.$$

• The internal energy of one mole is the sum of all the molecular kinetic energies $\frac{1}{2}mv^2$. It reads

$$\int_0^N \tfrac{1}{2}mv^2 \, dN = \tfrac{3}{2}RT.$$

We now appeal to the expression given above for ΔN in terms of A and μ; it leads to

$$\int_0^{+\infty} \tfrac{1}{2}mv^2 [(4\pi N A^3 v^2) \exp(-\mu^2 v^2)] \, dv = \tfrac{3}{2}RT.$$

This is simplified by taking the constant factors outside the integral, and by using the definition $k = R/N$ of Boltzmann's constant; one finds

$$4\pi A^3 m \int_0^{+\infty} v^4 \exp(-\mu^2 v^2) \, dv = 3kT.$$

Here we have another classic improper integral, whose value is $(3/8\mu^4)(\pi/\mu^2)^{1/2}$; this yields a second relation between A and μ:

$$mA^3 \frac{\pi^{3/2}}{2\mu^5} = kT.$$

To summarize: we now have two relations between A and μ:

$$A\frac{\pi^{1/2}}{\mu} = 1, \qquad mA^3 \frac{\pi^{3/2}}{2\mu^5} = kT.$$

They yield

$$\mu^2 = \frac{m}{2kT}, \qquad A = \left(\frac{m}{2\pi kT}\right)^{1/2}.$$

Consequently, Maxwell's velocity-distribution law reads

$$\rho(v)\,\Delta v = \frac{\Delta N}{N} = A^3[\exp(-\mu^2 v^2)]4\pi v^2\,\Delta v,$$

which after a brief calculation gives the end-result

$$\rho(v) = 4\pi \left(\frac{m}{2\pi kT}\right)^{3/2} \left[\exp\left(-\frac{mv^2}{2kT}\right)\right] v^2.$$

4 The velocity distribution for a perfect gas in two dimensions

This calculation is very like that in the preceding section, but it features only two variables, x and y, instead of the three variables x, y, and z. The resulting distribution is slightly different, and reads

$$\rho(v) = \frac{m}{kT} \left[\exp\left(-\frac{mv^2}{2kT}\right)\right] v.$$

It is plotted as the unbroken curve in Fig. 4.3. Notice in particular that it has finite slope at the origin, where the slope of the classic (three-dimensional) Maxwell distribution vanishes. Note also that the most probable speed, that is the position of the maximum of the distribution, now equals $(kT/m)^{1/2}$.

Next we must parametrize the distribution, so that theory can be compared with experiment.

If we have n molecules in all, then the number Δn of molecules in the range Δv is

$$\Delta n = n\,\frac{m}{kT} \left[\exp\left(-\frac{mv^2}{2kT}\right)\right] v\,\Delta v.$$

This becomes more manageable if we introduce two further parameters, namely

- the most probable speed $v_0 = (kT/m)^{1/2}$;

- the maximum value $(\Delta n)_0$ of Δn, which corresponds to $v = v_0$. Choosing $\Delta v = 1$, one obtains

$$(\Delta n)_0 = \frac{n}{v_0} e^{-1/2}.$$

Then the distribution reads

$$\Delta n = e^{1/2} \frac{(\Delta n)_0}{v_0} \left[\exp\left(-\frac{v^2}{2v_0^2} \right) \right] v \, \Delta v,$$

where $e^{1/2} \approx 1.65$.

The parameters v_0 and $(\Delta n)_0$ are the coordinates of the maximum of the experimental distribution constructed with $\Delta v = 1$. This is how the parameters can be determined most precisely.

Further reading

See Kittel (1969), Marcus and McFee (1959), Perrin (1920), and Rocard (1961) in the Bibliography.

Ludwig Boltzmann. Austrian physicist; born in Vienna, died at Duino, near Trieste (1844–1906). Inspired by Maxwell's kinetic theory of gases, he generalizes the method; his discovery of the canonical distribution in 1872 makes him the creator of statistical thermodynamics. Amongst his discoveries are the energy-equipartition theorem and the statistical expression for entropy (which opens the way to understanding thermodynamic irreversibility). As a convinced 'atomist' he is severely criticized by the 'energetists' Ostwald and Mach. In this conflict between statistical thermodynamics (very complicated) and classical thermodynamics (far simpler), common sense favours the latter. Only in 1908 do the experiments of Jean Perrin finally convince the physics community of the real existence of atoms. But Boltzmann is not alive to see it: deeply scarred by the attacks on him, he kills himself in 1906. Engraved on his tomb in a Vienna cemetery is the statistical expression for entropy. To this day it remains one of the fundamental formulae of physics. (Österreichische Nationalbibliothek)

5

Boltzmann, or probabilities as a matter of conviction (statistical physics)

canon   *a law, rule, edict; a general rule or axiom of any subject.*
canonical   *authoritative; orthodox; standard.*

Shorter Oxford English Dictionary

In developing the kinetic theory of gases, Maxwell showed that physical theory can fruitfully exploit probabilities that represent our ignorance. Boltzmann thought that this approach could be generalized, and was convinced that such generalization would open up fields of application far beyond the mere perfect gas. This, in 1872, was the birth of statistical physics, destined for further elaboration by Gibbs, though in rather more abstract fashion; that is why we confine ourselves here to Boltzmann's method, which has the advantage of reasoning directly in terms of the actual physical system.

5.1 Irreversible adiabatic expansion

Two metallic containers of equal volume are connected by a tube closed by a tap. Container A is filled with an ordinary gas like nitrogen, under a pressure of 1 bar, and container B is evacuated. The whole apparatus is immersed in a calorimeter at temperature T_i (Fig. 5.1).

We open the tap, and hear a whistling noise which indicates a rapid streaming of the gas through the connecting tube. Pressure drops in A and rises in B. Equilibrium is established in quite a short time, of the order of a few seconds at most; it is signalled by the cessation of the whistling, and by equal pressures, 0.5 bar, in the two containers. As regards the temperature of the calorimeter, it has not appreciably changed; in any case we can choose to consider the equality $T_f = T_i$ in such an expansion as a defining property of a

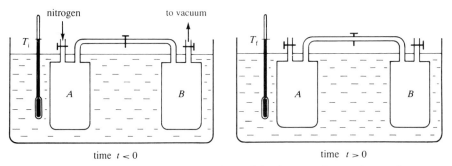

Fig. 5.1 Experimental realization of an adiabatic expansion at time $t = 0$.

perfect gas. This simple and very telling experiment† leads to three conclusions.

1. The *internal energy* of a perfect gas is independent of the pressure, because otherwise the temperature of the calorimeter would have changed.

2. The *equilibrium state* corresponds to equal pressures; moreover, since our two containers were chosen to have equal volumes, it corresponds to equal masses of gas in both.

3. This kind of expansion is *irreversible*: once the pressures have equalized, they remain equal indefinitely; one never observes the inverse process, which would result in a spontaneous accumulation of all the gas in *A*, leaving *B* empty. In other words, there is never a spontaneous return to the initial state.

What we have described is a macroscopic process, though we know that on the microscopic scale a gas consists of very many molecules. Can one explain the observed laboratory-scale irreversibility in terms of the molecular nature of the gas? The attempt leads one to look for a simple model of the process.

5.2 A model for irreversible adiabatic expansion

The model, though simple, is perfectly true to the essential physics of the process. The two containers (which have three-dimensional volumes) are represented by two flat regions (which have two-dimensional areas). The

---

† The experiment, when performed with real gases, is amply sufficient to suggest certain limiting properties that are then attributed to perfect gases, as listed in the three conclusions that follow.

model, drawn in Fig. 5.2, is a rectangle divided into two regions A and B by a partition represented by the broken line. The molecules are pictured as discs that are confined to region A as long as the partition persists. The partition is suddenly removed at time $t = 0$, mimicking the opening of the tap; thereafter the molecules are free to range through both regions, and they distribute themselves over the entire surface open to them. Equilibrium is established very fast, with the same mean number of molecules and therefore with the same mass of gas in each region; such equilibrium is statistical, which means that the actual numbers fluctuate slightly around their mean values. This model gives a satisfactory qualitative description of the expansion, but naturally we need more than that. The challenge is to give a quantitative account of the phenomena; and, since we are taking one step at a time, we start with just 20 molecules.

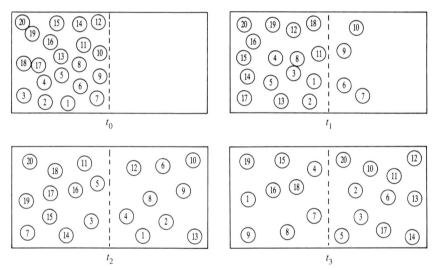

Fig. 5.2 Typical stages of an adiabatic expansion, as modelled on a computer, for 20 molecules.

Since the advent of computers it would be possible, given 20 molecules, to work with a *deterministic theory*, even though the calculations would be quite heavy; the pioneers in this field are J. B. Adler and T. E. Wainwright, who have managed to deal with as many as 100 molecules [1]. At that time (in 1959) a very large computer was needed; today the problem is accessible with a good micro.

The basis of the calculation is just the basis of deterministic theory generally, namely rational mechanics.

1. At time $t=0$ every molecule is assigned a position (x, y) and a velocity (v_x, v_y); these are the initial conditions.

2. Every molecule moves in a straight line, but undergoes an elastic collision on meeting either another molecule or the wall; apart from these collisions the molecule experiences no interactions.

3. Every molecule is individually recognizable, since otherwise no deterministic calculation would be possible; in other words we assume that the molecules are distinguishable, and label each with a number.

Whatever program one uses, and whatever initial conditions are chosen, the general features of the results are always the same, being like those illustrated in Fig. 5.2:

- for $t<0$ the molecules remain confined to region A;
- at time $t=0$ the partition is removed;
- at time $t=t_1$ the molecules start to spread into region B;
- at time $t=t_2$ the molecules are distributed almost equally between A and B;
- at time $t=t_3$ this equal (and equilibrium) distribution is maintained on average, even though exchanges of molecules continue between the two regions.

Thus, deterministic theory applied to just a few molecules affords an explanation of the irreversibility of such expansions. But we must stress that the theory takes very long to implement, both in writing and in running the program.

The *probabilistic theory* for 20 molecules works incomparably faster. One does not here deny that the phenomenon is basically deterministic, but one recognizes that it takes too long to exploit this fact, given that the object is not to describe the expansion in detail, but only to explain why it is irreversible when viewed macroscopically. Boltzmann bases his analysis on the following observation: the molecules are so fast, and their collisions so frequent, that the system rapidly loses or at least appears to lose track of the initial conditions. Typically this leads us into the realm of probabilities through ignorance, and we introduce the probabilistic theory through two assumptions.

1. Every molecule has equal *a priori* probabilities of being in region A and in region B. This assumption becomes intuitively plausible as soon as we admit that the position of a molecule is governed by chance, which puts us firmly on the level of a game of heads or tails featuring equal probabilities: $a=b=\frac{1}{2}$ (Section 2.4).

2. The system evolves spontaneously from the less towards the most probable

state. This is the assumption, likewise intuitive, that will give us the key to irreversibility.

Our next task is to exploit these two assumptions through a game of heads or tails, with the sole difference that instead of playing twenty successive games with the same coin, we play just one game, consisting of the observation of the twenty molecules at time t. The results are the same, in the sense that both are described by the binomial distribution (Section 2.4).

Envisage therefore the system at some given time $t > 0$, late enough after the removal of the partition to validate recourse to chance; such times correspond to t_2 and t_3 in Fig. 5.1. Our aim is to determine the probability to be assigned to any state of the system. Boltzmann distinguishes between macro-states and micro-states.

- A *macro-state* is specified by the number of molecules in each region: for instance, there might be $m_A = 5$ molecules in A and $m_R = 15$ molecules in B. One always has $m_A + m_B = 20$, whence there are 21 macro-states of the system.

- A *micro-state* is defined by the individual numerical labels of the molecules present in each region: for instance, molecules number 2, 3, 9, 12, 14 might be in region A, and molecules number 1, 4, 5, 6, 7, 8, 10, 11, 13, 15, 16, 17, 18, 19, 20 in region B. There are $2^{20} = 1\,048\,576$ micro-states in all, and the probability assigned to each such state is $2^{-20} = 1/1\,048\,576$.

Boltzmann notes that very many different micro-states give the same macro-state. For example, the macro-state $(m_A = 5, m_B = 15)$ can be realized through any combination of five numbered molecules chosen from a set of 20 numbered molecules. The number of such combinations is $\binom{20}{5} = 15\,504$.

Finally, one can calculate the probability assigned to the macro-state $(m_A = 5, m_B = 15)$ from the formula $P(5) = \binom{20}{5} 2^{-20} = 0.014\,786$.

We can in this way calculate the probabilities $P(m_A)$ assigned to the 21 possible macro-states (see Appendix 1), and display the results graphically (Fig. 5.3a), which leads us to three important observations.

1. The most probable macro-state corresponds to $m_A = \bar{m} = 10$, that is to equal numbers of molecules in the two regions.† Its probability is $0.176\,197$, which is quite high.

2. The least probable macro-state corresponds to $m_A = 0$ or to $m_A = 20$, that is to an accumulation of all the molecules in the same region. Its probability is $0.000\,001$, which is very low.

3. There are fluctuations around the median value $m_A = 10$; they are almost wholly straddled by the interval $\pm 3\sigma$, that is by ± 6.7. Consequently a

---

† The most propable value of m_A equals the mean $\bar{m}$ because the distribution is symmetric.

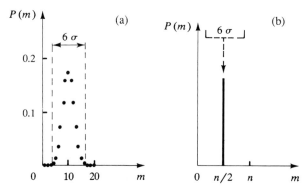

Fig. 5.3 Statistical study of the equilibrium reached after an adiabatic expansion:
(a) the distribution for 20 molecules; (b) the distribution for 10^{23} molecules.

spontaneous return of the system to macro-states outside this interval [22]
is most unlikely.

We conclude that the system has evolved spontaneously from the initial
situation $m_A = 20$ to a statistical equilibrium state near $m_A = 10$. For all
practical purposes one can exclude the possibility of any return to the original
state: accordingly, the transformation is irreversible.

Under normal conditions a set of 10^{23} molecules in the gas phase occupy a
volume of the order of 4 litres, close enough to the expansion in the
demonstration experiment described in the preceding section. Can we devise a
model for this situation too?

With a deterministic theory the answer is certainly no, because one would
now have to deal with 10^{23} molecules instead of a mere 20. Technically the
calculations would be utterly impossible, as already explained in the preceding
chapter (Section 4.1), in the course of our search for the velocity distribution.

By contrast, a probabilistic theory inspired by our study of 20 molecules
delivers the results at once. The various possibilities could be evaluated
through the formula

$$\binom{n}{m_A} = \frac{n!}{m_A! m_B!},$$

but actually this is unnecessary, because we can appeal directly to general
properties of the binomial distribution, as shown in Fig. 5.3b.

- The most probable macro-state corresponds to $m_A = \bar{m} = \frac{1}{2}n = 0.5 \times 10^{23}$.

- The least probable macro-state corresponds to $m_A = 0$ or to $m_A = n$.

- The order of magnitude of the fluctuations is governed by the standard deviation $\sigma = (nab)^{1/2} = 1.6 \times 10^{11}$. Almost all are straddled by the interval $\pm 3\sigma$, that is by $\pm 4.7 \times 10^{11}$. Relative to the mean value $\bar{m} = 0.5 \times 10^{23}$, this interval is extremely narrow:

$$\frac{3\sigma}{\bar{m}} = \frac{4.7 \times 10^{11}}{0.5 \times 10^{23}} \approx 10^{-11}.$$

That is why in Fig. 5.3b the distribution appears so sharply peaked, with no visible width at all. Statistical equilibrium is established around $m_A = \bar{m} = \frac{1}{2}n$, with very high precision: no pressure-measuring apparatus whatever has the remotest chance of detecting any deviations.

We deduce that once the partition is removed *the system evolves irreversibly*, from the initial state $m_A = n$, where all the molecules are in region A, to a state of statistical equilibrium around $m_A = \frac{1}{2}n$, where the molecules are distributed equally between the two regions A and B. A spontaneous return of all the molecules to region A is not *a priori* impossible; but it is highly improbable by everyday human standards, since one would need to wait for a time much longer than the age of the Universe in order to have even a small chance of witnessing it.

Given our theoretical picture of irreversible adiabatic expansion, three largely philosophical problems come to mind quite naturally.

1. The deterministic nature of molecular motions is admitted in principle, even though they are treated probabilistically. The laws of classical mechanics, which are reversible ('invariant under time-reversal'), are taken as valid on the microscopic scale; nevertheless they generate, on the macroscopic scale, a law prescribing behaviour that is irreversible. How come? One can understand this better after analysing the expansion as simulated for 20 molecules on a computer (Fig. 5.2). If we let time run backwards, starting from t_3, then we do observe a spontaneous return of the molecules to region A at time 0. This reasoning is as valid for 10^{23} molecules as for 20, and does demonstrate at least the possibility of spontaneous accumulation in A. Why then is this never observed, nor any other process like it? The answer to this question came some decades later, with Henri Poincaré's recurrence theorem: one can always find a time lapse, appropriately called the *recurrence time*, after which the system will come as close as one pleases to its initial state. In other words, after opening the tap in our expansion experiment, we shall observe spontaneous expansion followed by spontaneous recompression into container A; but the recurrence time is very long, much longer than the age of the Universe. Hence the recurrence theorem is irrelevant in practice, though very important conceptually. To summarize, we assert, with Boltzmann, that spontaneous accumulation in container A is possible but very unlikely; we also assert, with

Poincaré, that it is certain but very far in the future; and there is no contradiction between the two assertions.

2. The experiment is reproducible and gives identically the same result every time; therefore it is governed by laws that are deterministic. Nevertheless, we have explained it in terms of random molecular motions, reasoning within a framework of probabilities through ignorance. We conclude that chance operating on the microscopic level can generate determinism on the macroscopic level. In any analysis of the complex relationship between chance and determinism, this is a most interesting observation.

3. In order to construct the theory governing a deterministic system of 10^{23} molecules, we have exploited probabilities through ignorance, and with complete success. The reason for this success deserves discussion. The Soviet physicist Lev Landau has shown that a classical system requiring infinitely many parameters would behave in a totally random fashion; in other words it would be random unavoidably, and not merely by reason of our ignorance. Our system of 10^{23} molecules depends on 6×10^{23} parameters; this number though not infinite is very large, and can be considered as infinite for practical purposes. We have in this way been led from chance by reason of ignorance to chance unavoidable, which is governed by the laws of probability discussed in Chapter 2. That is why we have struck so lucky with our appeal to the binomial distribution; and why the use of probabilities through ignorance, far from being a mere second-best reflecting the limits of our numerical capabilities, in fact takes us very close to the laws of nature.

5.3 Phase space

We have been considering a gas distributed over two regions, with every molecule occupying either region at random. Boltzmann is convinced that this notion can be generalized, provided we extend our concept of what constitutes a 'region'. In order to describe the properties of a given mass of gas in ordinary, three-dimensional, physical space, he introduces an abstract six-dimensional space, where the position of every molecule is specified by its three ordinary position coordinates q_x, q_y, q_z, plus the three Cartesian components of its momentum p_x, p_y, p_z.† Boltzmann imagines further that phase space is divided into cells of equal volume, and that every molecule has the same *a*

---

† The most general definition of phase space features the coordinates q_x, q_y, q_z, plus their time-derivatives $\dot{q}_x$, $\dot{q}_y$, $\dot{q}_z$. However, use of the momenta $p = m\dot{q}$ instead of the velocities $\dot{q}$ (to which they are proportional) will enable us to determine the dimensions of the elementary cells of phase space by appeal to the Heisenberg inequalities (see the rest of the present section).

priori probability of occupying any one of these cells; thus he introduces concepts of *quantization*, almost thirty years before Planck.† The actual volume of a cell is unimportant to the argument that follows, though we do, today, know what it is. We start with just two axes q and p (Fig. 5.4).

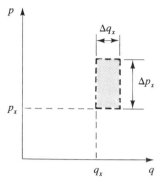

Fig. 5.4 Quantization of two-dimensional phase space by subdivision into cells of area $\Delta q_x \Delta p_x \geq h$.

We know from quantum mechanics that if one tries to measure q_x and p_x simultaneously, then the results are unavoidably scattered through ranges Δq_x for q_x and Δp_x for p_x respectively, and that these ranges are related by the Heisenberg inequality

$$\Delta q_x \Delta p_x \geq h,$$

where h is Planck's constant ($h = 6.626\,176 \times 10^{-34}$ J s).

Accordingly, we can assign to each molecule a rectangular 'occupancy region' in the (q_x, p_x)-plane, shown shaded in Fig. 5.4; the area of this region cannot be less than h. Next, we extend the argument to the full set of six coordinates $q_x, q_y, q_z, p_x, p_y, p_z$; this puts us into six-dimensional phase-space, with elementary cells that are likewise six-dimensional. We can no longer draw them, but we can still calculate their volume, which is at least h^3.

We shall, accordingly, keep in mind that in statistical physics the point representing a molecule does not describe a continuous trajectory through phase space; rather it progresses by jumping from one elementary cell into another‡ Thus it becomes natural to ask how the molecules are distributed over the different cells.

---

† Boltzmann never gave a precise specification of these cells; what he stressed was that molecules exist, and the possibility of assigning them to sub-regions of phase space. As regards phase space itself, it was invented by Maxwell and elaborated by Gibbs.

‡ We have sidled somewhat disingenuously from the classical cells of Boltzmann to quantum cells governed by the Heisenberg inequalities, that is from probabilities through ignorance to probabilities inherent in nature. This remark will become clearer automatically in Chapters 7 and 8; meanwhile we note that there was of course no reason in Boltzmann's day for not sticking to chance through ignorance, given the state of development of physics in the nineteenth century.

5.4 The Boltzmann distribution

Consider n molecules of a perfect gas; n is large, and we think of it as being of the order of 10^{23}. Each of these molecules occupies one of the many cells of phase space, at random. There are far too many cells for them to be labelled by the letters of the alphabet, as they were in our model of irreversible expansions. Hence we label every cell by an index i, stipulating only that the number of cells be very large, yet much smaller than the number of molecules. This condition is necessary in order to validate the approximations required by the calculation; just because it is not intuitively obvious, it illustrates Boltzmann's insight, both physical and mathematical. The probability that a molecule picked at random occupies cell i is a constant independent of i. By contrast, the energy ε_i of a molecule in cell i does depend on the cell, and thereby on i. Finally, the numbers of molecules in the different cells are denoted by the sequence of occupation numbers $m_1, m_2, \ldots, m_i, \ldots$

Figure 5.5 sketches the basic principle of Boltzmann's calculation. As before, we must distinguish between two types of states.

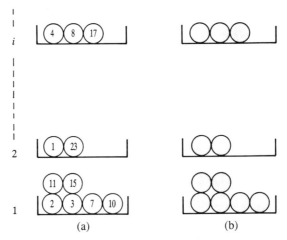

Fig. 5.5 The basic elements in the derivation of the Boltzmann distribution: (a) a micro-state (the order within a given cell is irrelevant); (b) a macro-state.

- *Micro-states* are defined by listing every molecule by its serial number and the number of its cell. Boltzmann called such states 'complexions', but nowadays one prefers to speak of *accessible states*.

- *Macro-states* are defined merely by specifying the number of molecules in each cell, that is by the sequence of occupation numbers $m_1, m_2, \ldots, m_i, \ldots$

We aim to determine the properties of the most probable state, which as we already know corresponds to statistical equilibrium. Hence we look for the number of micro-states that correspond to a given macro-state. In our previous system with just two cells A and B, this was the number of combinations

$$\binom{m}{n} = \frac{n!}{m_A! \, m_B!}.$$

In the general case now under study, the requisite number is given by a generalization of this formula, namely by

$$\Omega = \frac{n!}{m_1! \, m_2! \ldots m_i! \ldots}, \quad \text{or better by } \Omega = \frac{n!}{\prod_i m_i!}.$$

Here Ω is the number of accessible micro-states corresponding to the prescribed macro-state.

Two further conditions come into play.

1. The sum of all the m_i equals the total number n of molecules: $\sum_i m_i = n$.

2. The sum of the energies of all the molecules equals the internal energy of the gas: $\sum_i \varepsilon_i m_i = U$. The energy U is constant in time, because the system is isolated.

To calculate the properties of the system in statistical equilibrium amounts to determining the maximum of Ω (see Appendix 3). The result is expressed very simply through the Boltzmann distribution, also called *the canonical distribution*:

$$P(\varepsilon_i) = \frac{\exp(-\varepsilon_i/kT)}{Z},$$

where $P(\varepsilon_i)$ is the probability that any given molecule occupies a given state having energy ε_i, k is Boltzmann's constant, T is the absolute temperature in kelvin, and $Z = \sum_i \exp(-\varepsilon_i/kT)$ is a function called the 'partition function'.

By the same token we can exhibit the statistical nature of entropy. Given a physical system consisting of two parts 1 and 2, having entropies S_1 and S_2 respectively, we know from classical thermodynamics that the entropy of the joint system is the sum $S = S_1 + S_2$. But if we reason in terms of micro-states, we see that there are Ω_1 such states for part 1 and Ω_2 for part 2; the combination rule then yields the total number of micro-states for the joint system $(1 + 2)$ as $\Omega = \Omega_1 \Omega_2$. This correspondence between product and sum led Boltzmann to a further insight: since, mathematically, what turns products into sums is the logarithm, he was led to posit proportionality between S and $\ln \Omega$, thus giving a statistical definition of entropy; it reads

$$S = k \ln \Omega,$$

where S is the entropy of the system in statistical equilibrium, k is Boltzmann's constant, and Ω is the number of accessible micro-states. The value $k = R/N$ of the proportionality constant emerged only from a long analysis that came later [14].

The concrete physical significance of the canonical distribution is illustrated step by step in Fig. 5.6, for the very simple case where the energy levels are equally spaced and non-degenerate.†

- The graph in Fig. 5.6a shows the probability $P(\varepsilon)$ as a function of ε; it is a decreasing exponential, as prescribed by the Boltzmann distribution. The probability $P(\varepsilon)$ is measured in units of Z, and ε in units of kT.

- Figure 5.6b shows a more tangible result, corresponding to n molecules; in this case the mean occupancy of state ε is $\bar{m}(\varepsilon) = nP(\varepsilon)$, and we indicate, as a histogram, the values $\bar{m}(\varepsilon)$ as functions of ε. This kind of histogram, plotting ε horizontally and $m(\varepsilon)$ vertically, will occur again in the simulation of Einstein's model of a solid, in Section 5.8.

- Figure 5.6c shows the function inverse to the preceding one, that is it plots $\bar{m}(\varepsilon)$ horizontally and ε vertically. Here the energy levels feature in the usual way as the rungs of a ladder; we have drawn the molecules as sitting on these rungs, in order to illustrate the significance of $\bar{m}(\varepsilon)$. This is a good way of seeing that the molecules tend to accumulate preferentially in the states of low energy, with fewer and fewer per state as the energy rises.

The *law of increasing entropy*, which governs all irreversible processes, is a natural consequence of the statistical expression $S = k \ln \Omega$. In our irreversible expansion for instance, the opening of the tap increases the number of states accessible to the gas molecules; and if Ω rises, then, necessarily, the entropy S rises too. From our model we can calculate that the increase is $\Delta S = kn \ln 2$, where n is the total number of molecules. The same result can be exhibited in terms of amounts of information (Fig. 5.7). Before the partition is removed, we can guarantee that every molecule is located between 0 and 10 cm. Afterwards, our guarantee is less precise, warranting locations only between 0 and 20 cm. Accordingly we have a loss of information, which corresponds to a rise in entropy.

---

† 'Equal spacing' means that successive energies ε_i are proportional to successive integers. 'Non-degenerate' means that to every energy ε_i there corresponds just one state i in phase space; in that case we can dispense with the index i, and the energy ε serves to label both the level and the state.

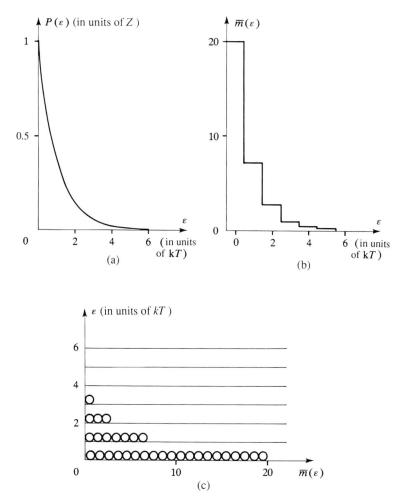

Fig. 5.6 Plots of Boltzmann distributions: (a) universal curve using normed units; (b) histogram indicating the mean occupancies $\bar{m}(\varepsilon)$; (c) distribution of the molecules over energy levels.

5.5 The barometer formula

From the Boltzmann distribution applied to a perfect gas we can recover, with only a minimal amount of calculation, the Maxwell velocity distribution already established in Chapter 4:

$$p(v) = 4\pi \left(\frac{m}{2\pi kT} \right)^{3/2} \left[\exp\left(-\frac{mv^2}{2kT} \right) \right] v^2.$$

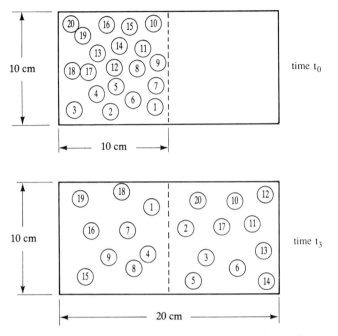

Fig. 5.7 Rise in entropy corresponds to loss of information.

But the Boltzmann distribution tells us more, allowing us to deal also with a perfect gas in a gravitational field. For this one need only note that the exponential in the Maxwell distribution stems directly from the exponential in the Boltzmann distribution: in general, the argument of the exponential is $-\varepsilon/kT$.

- Without gravity, the mechanical energy of a molecule is purely kinetic; for a perfect monatomic gas it reads $\varepsilon = \frac{1}{2}mv^2$.

- In a gravitational field, the mechanical energy is the sum of kinetic and potential energies; hence it reads $\varepsilon = \frac{1}{2}mv^2 + mgz$, and the distribution depends on two variables, v and z.

However, if we integrate the probability density ρ over all values of v from 0 to $+\infty$, and if we take z, g, and T as constants, then we obtain the very simple result

$$\rho(z) = \rho(0) \exp\left(-\frac{mgz}{kT}\right).$$

This identifies $\rho(z)\Delta z$ as the probability of finding a given molecule between the heights z and $z + \Delta z$, irrespective of its velocity.

By progressing through a sequence of physical quantities all proportional to each other, we can now derive the pressure at height z. The pressure is proportional to the number of impacts per second on the wall of the manometer; the number of impacts is itself proportional to the number of molecules per unit volume at height z; finally this number density is proportional to the probability density $\rho(z)$.

Thus we derive the barometer formula for an isothermal atmosphere in a uniform gravitational field; it reads

$$P(z) = P(0) \exp\left(-\frac{mgz}{kT}\right),$$

where $P(z)$ is the pressure at height z; $P(0)$ is the presssure at height 0; m is the mass of one molecule; g is the acceleration due to gravity; z is the height; k is Boltzmann's constant; and T is the absolute temperature.

The formula is given here in terms of the microscopic quantities m and k, but it can be rewritten in terms of macroscopic quantities by multiplying through by Avogadro's number. Then it features the molar mass $M = mN$ and the perfect-gas constant $R = kN$. Thus, the barometer formula in macroscopic terms reads

$$P(z) = P(0) \exp\left(-\frac{Mgz}{RT}\right),$$

where M is the molar mass (in kilograms) and R is the perfect-gas constant.

This macroscopic law was known before Boltzmann, originally from experimental and theoretical studies of the statics of fluids. Thus the result derived microscopically is compatible with physics known beforehand. It remains to demonstrate that the canonical distribution has many other uses, far beyond the case of a perfect gas in a gravitational field. This was to be shown by Albert Einstein in 1905, with his theory of the specific heat of solids; by Paul Langevin in 1907, with his theory of paramagnetism; and by Jean Perrin in 1908, with his experimental study of dust suspensions in liquids.

The barometer formula can serve as a reasonable model for the Earth's atmosphere, yielding the variation of pressure with height.[†] This is plotted in Fig. 5.8, where P is calculated using the molar mass of nitrogen,[‡] which is the main constituent of air (79% by volume), and an absolute temperature of $T = 300$ K. The graph shows, in particular, that a measurement of the

---

† Strictly speaking one should take account of the drop in temperature at high altitude, which could be done by adding a correction term to the exponent. But the constant-temperature approximation suffices to exhibit the gist of the matter, and is what we show in Fig. 5.8.

‡ The isothermal approximation is not very good, which is why we settle simply for nitrogen, with $M = 28$. One could, of course, do better by defining a 'mean molar mass' for air, with $M = 29$.

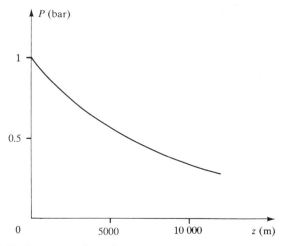

Fig. 5.8 The barometer formula for pressure P as a function of height z.

atmospheric pressure P determines the altitude z; this is the principle of the altimeter, used by climbers and fliers.

5.6 Simulation of the Earth's atmosphere

As in Chapter 4, we proceed to illustrate the Boltzmann distribution by simulation on a laboratory scale. Again we use an air table. But, in order to exhibit the effects of gravity, we now incline the table-top at a small angle α to the horizontal (Fig. 5.9). This yields an excellent two-dimensional simulation of the terrestrial atmosphere, where every molecule, represented by a hard disc, is under the permanent influence of the component $g \sin \alpha$ of the gravitational acceleration, and is under the influence also of a random sequence of elastic collisions. The theory is very similar to that for a three-dimensional gas, and leads to a result of much the same kind:

$$\rho(z) = \rho(0) \exp\left(-\frac{mgz \sin \alpha}{kT}\right).$$

By subdividing the table-top into horizontal strips of width a and 'height' $\Delta z = 1$, we can identify two variables proportional to $\rho(z)$ and $\rho(0)$ respectively; they are

• the number Δn_z of discs in the strip of width $\Delta z = 1$ at height z;

- the number Δn_0 of discs in the strip of width $\Delta z = 1$ at height 0.

The equation can then be written very simply, whether in exponential or in logarithmic form:†

$$\Delta n_z = \Delta n_0 \exp\left(-\frac{\sin \alpha}{H} z\right), \qquad \ln \Delta n_z = \ln \Delta n_0 - \frac{\sin \alpha}{H} z.$$

The second equation results from a transformation of the first, a transformation that allows the data to be analysed more conveniently.

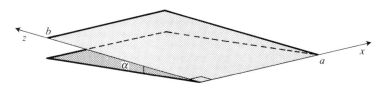

Fig. 5.9 Sketch of the sloping table-top, indicating the choice of axes.

The experiment is done with 25 discs moving on an air table inclined at an angle α to the horizontal, with $\alpha \approx 0.5°$. The inclination and the speed of agitation of the frame are chosen by compromise: the molecules must not reach the top edge of the frame, because this would affect the distribution. Steady agitation eventually establishes a steady state, with the density decreasing from bottom to top. One takes snapshots of the table using flash; for analysis each picture is divided into 12 horizontal strips. The height of each strip is unity by definition, $\Delta z = 1$, and we count the number Δn of molecules in each strip.

The table below shows the results from analysing 28 exposures, featuring 700 molecules in all.

Strip number	0	1	2	3	4	5	6	7	8	9	10	11
Δn	112	135	112	108	81	51	40	23	21	15	2	0

† For simplicity we have written $1/H = mg/kT$, side-stepping any need to define the 'temperature of the discs', which would merely be a distraction.

The numbers $\Delta n \pm (\Delta n)^{1/2}$ as functions of z are plotted in Fig. 5.10. The experimental points do appear to approximate to a curve that looks exponential, but for a more searching test one adopts semi-logarithmic coordinates. Then one asks how closely the data conform to the relation in its transformed version, that is whether they fall close to a straight line whose equation reads

$$\ln \Delta n_z = \ln \Delta n_0 - \frac{\sin \alpha}{H} z.$$

Out of the 12 data points, seven are aligned very well. The other five lie off the line, and it is interesting to ask why.

1. Two points at high altitude (strips 11 and 12) correspond to measurements whose precision is very low because of poor statistics ($\Delta n = 2$ and $\Delta n = 0$). Their deviation from the straight line is irrelevant.

2. The three points from the lowest altitudes (strips 1 to 3) are also shifted; but one must realize that they correspond to low values of z where the 'molecules' are very crowded, while for a perfect gas it is assumed that they are far apart relative to their diameters. We have thus departed from the simplicity of the model, and it is to be expected that these points should depart from the straight line given by the elementary theory.

If we disregard these five points, we are left with the central region, where the agreement between theory and simulation is perfectly acceptable in the quantitative sense of the words. But the real purpose of the simulation is to facilitate qualitative analysis and understanding. With a movie-camera whose axis is normal to the table-top, one can record the random motion of the discs: the film shown to a large audience always draws the same reaction, to the effect that only the moving image affords any real understanding of what is meant by the thermal agitation of the molecules in the Earth's atmosphere.

5.7 The Einstein solid

A simple solid is modelled by a collection of atoms arranged regularly. This model accounts for the geometry of the cleavage planes, and also for the optical properties; but thermodynamic properties are left open. To specify them, we imagine that every atom can oscillate about its lattice site, and that the amplitude of such oscillations increases with the temperature. Accordingly, each atom is modelled as a sphere linked to its neighbours by six springs. As a first approximation the neighbour atoms are taken as fixed, only the central atom being free to oscillate (Fig. 5.11). The three directions for these

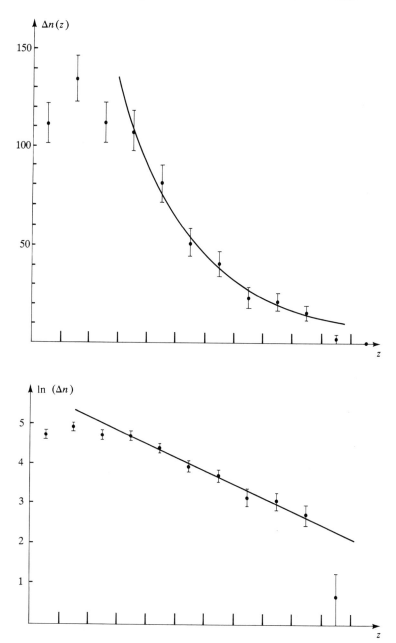

Fig. 5.10 Measured densities as functions of height z. Top: $\Delta n = f(z)$; the distribution is exponential. Bottom: $\ln \Delta n(z) = g(z)$ shown in semi-logarithmic coordinates. The measured points lie on a straight line, except near the ends. (These deviations are discussed in the text.)

vibrations are mutually orthogonal; hence one can say that we have assigned to every lattice site three independent but dynamically identical harmonic oscillators.

The classical theory[†] using Boltzmann's statistical mechanics is easily applied. Denote the three directions of vibration as x, y, z, and write the mechanical energy corresponding say to the variable x as

$$E = \tfrac{1}{2}mv_x^2 + \tfrac{1}{2}m\omega^2 x^2,$$

where $\tfrac{1}{2}mv_x^2$ is the kinetic energy; $\tfrac{1}{2}m\omega^2 x^2$ is the potential energy; v_x is the velocity component along x; m is the mass of the atom; and $\omega/2\pi$ is the frequency, common to all the oscillators.

Next we must introduce a theorem due to Boltzmann, which proves extremely useful in statistical physics, namely the energy-equipartition theorem [14]:

If the energy of a physical system can be written as a sum of mutually independent quadratic terms, then the mean energy associated with each of these terms is $\tfrac{1}{2}kT$.

The application to an oscillator in our crystal lattice is immediate, because its energy is indeed the sum of (two) independent quadratic terms; hence the mean energy of thermal agitation is $\bar{\varepsilon} = kT$ per oscillator.

For the corresponding macroscopic result, we first multiply by 3 to go from oscillator to atom; and then by Avogadro's number to go from one atom to one mole, using $kN = R$. This yields the mean energy $\bar{E}$, that is the internal energy U of one mole: $\bar{E} = U = 3RT$.

Finally, the molar specific heat at constant volume,[‡] C_V, follows trivially on differentiation with respect to T:

$$C_V = dU/dT = 3R = 25 \text{ J K}^{-1} \text{ mol}^{-1}.$$

At room temperatures, the molar specific heat C_V is a constant, equal to $25 \text{ J K}^{-1} \text{ mol}^{-1}$ for all simple solids.

This result, known as Dulong and Petit's law, had already been discovered empirically in systematic calorimetric measurements on solids. It is a major success of Boltzmann's statistical physics to have explained it in terms of the oscillations of atoms, all the more so because the model used is so simple. To illustrate the success we quote some calorimetric data on C_V at 25°C.

From aluminium to lead, theory and experiment agree very well, given the crudity of the model used for the oscillations. However, at low temperatures the specific heat C_V is no longer constant; it drops with the temperature, and

---

† 'Classical' in this context means using continuous variables and working at temperatures that correspond to the 'classical limit' of quantum mechanics; generally this applies at room temperatures.

‡ The thermal expansion of solids is negligibly small: hence $C_P = C_V$.

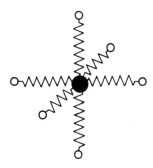

Fig. 5.11 Model for the vibrations of an atom in a crystal.

Simple solid	Al	Ag	Cu	Sn	Pb
$C_V(\text{J K}^{-1}\,\text{mol}^{-1})$	23.4	24.4	23.8	25.4	24.8

vanishes at the absolute zero. Thus the classical model proves inadequate, and it is a quantum model, proposed by Einstein in 1905, that resolves the difficulty.

The quantum theory of crystalline solids uses the same model of oscillating atoms, that is the same set of three dynamically identical but independent harmonic oscillators (Fig. 5.11), but now the energies possible for each oscillator are given by the quantal formula

$$\varepsilon = (v + \tfrac{1}{2})\frac{h\omega}{2\pi},$$

where v is a non-negative integer and $\omega/2\pi$ is the frequency of the oscillator. The product $h\omega/2\pi$ is written $\hbar\omega$, where $\hbar = h/2\pi$. Einstein showed that the product is equal to the quantum of energy appropriate to a harmonic oscillator of frequency $\omega/2\pi$.

The probability that an oscillator is in the state with energy ε is given by the Boltzmann distribution:

$$P(\varepsilon) = \frac{1}{Z}\exp\left(-\frac{\varepsilon}{kT}\right).$$

From these two relations one can calculate the internal energy U per mole, and one finds [12]

$$U = 3N\hbar\omega \left(\tfrac{1}{2} + \frac{1}{\exp(\hbar\omega/kT) - 1} \right).$$

Differentiating U with respect to T we obtain C_V, the molar specific heat at constant volume:

$$C_V = \frac{dU}{dT} = 3R \left(\frac{\hbar\omega}{kT} \right)^2 \frac{\exp(\hbar\omega/kT)}{[\exp(\hbar\omega/kT) - 1]^2}.$$

Crucial to this relation is the ratio of $\hbar\omega$, which is the elementary quantum of excitation, to kT, which defines the energy scale natural to statistical physics.

- If $\hbar\omega \ll kT$, then we are at the 'classical limit' of quantum mechanics, which is appropriate at room temperatures. It is easily seen that C_V simplifies, reducing in the limit to $C_V = 3R$. We have thus recovered Dulong and Petit's law, which corresponds to the classical theory.

- If $\hbar\omega \gg kT$, then quantum effects dominate, and it is easy to see that the specific heat C_V vanishes as the absolute temperature tends to zero.

Einstein's model of solids promotes Boltzmann's statistical physics from the explanatory to the predictive, in that

(a) it explains a property already known, i.e. Dulong and Petit's law;

(b) it predicts something new, unknown to Dulong and Petit, i.e. the rule that specific heats decrease as the temperature falls towards absolute zero.

This predicted decrease is verified by experiment, which is an endorsement of Boltzmann's statistical physics, and final confirmation of its status as a true scientific theory.†

5.8 Simulation of the Einstein solid

In our present example, it is a particular challenge to devise an illustrative macroscopic system obeying the same laws as the microscopic system under consideration, because quantization is very much peculiar to atoms and molecules. Fortunately the computer is flexible enough to tackle any problem, classical or quantum. Its speed allows it to deal with very large numbers of events, which improves the statistics. Finally, printers and screens afford immediate visibility, making the results vivid and tangible. Thus one can

---

† It should be added that Debye in 1912 proposed a model for solids more elaborate than Einstein's, introducing the concept of collective vibrations, and replacing the single vibration frequency by a continuous frequency spectrum with quite a wide range [12].

explore in succession the influence of the various physical parameters of the model; in effect one does experiments on the computer, which becomes a new tool for teaching statistical physics.

We have used a microcomputer to simulate an Einstein solid. As suggested by Black and Ogborn in 1976, we adopt two simplifications: the three-dimensional crystal is replaced by a plane array subdivided into cells, each representing one lattice site (Fig. 5.12); and each site contains only one oscillator. These simplifications are acceptable for our purposes, because they leave the general features of the results unaffected.

The energy of each (quantized) oscillator is an integer multiple of $\hbar\omega$. Thermal interaction between them consists in the transfer of quanta $\hbar\omega$. The number of points assigned to each cell is the number of quanta possessed by the oscillator. Transfer of quanta are governed by a lottery, which we call the temperature game.

Figure 5.12 shows the start of the game. The initial state (Fig. 5.12a) assigns one quantum to every cell. The first step consists in drawing four numbers at random: in this case they are 4, 2, 3, 5. The first two, $x_1 = 4$ and $y_1 = 2$, select the coordinates of the cell which is to lose one quantum. The last two, $x_2 = 3$ and $y_2 = 5$, select the cell to which this quantum is transferred (Fig. 5.12b). There are two further rules:

- if the initial cell is empty, the draw is disregarded;
- if, in the course of the game, the second cell of a draw is the same as the first, then that draw too is disregarded.

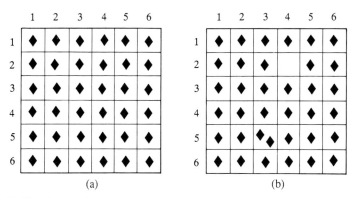

(a)   (b)

Fig. 5.12 Simulation of an Einstein solid with a 6×6 matrix. Exchanges at random, governed by throwing dice: (a) the matrix in its initial state; (b) the matrix after the random transfer of one quantum between two cells.

For a 6×6 array the temperature game can be run with dice, but the approach to equilibrium is very slow. With larger arrays dice are inconvenient, and a microcomputer becomes essential; the function RND produces a sequence of pseudo-random numbers on request, mimicking a game of chance for all practical purposes. The software supplies the results in the form of coordinates (x, y), and updates the occupancies of the cells after every step. Figure 5.13 shows stages of this game for a 16×16 matrix. Initially, every cell contains one quantum (Fig. 5.13a). After 20 000 steps† thermal equilibrium is fully established (Fig. 13b). One might have thought that such equilibrium would correspond to a uniform distribution of one quantum per cell, as in the initial state. But in fact this is far from being the case, because many cells are empty, while others contain several quanta. The results are shown in the table and on the histogram, giving the number m of cells as a function of the number v of quanta they contain (Fig. 5.14).

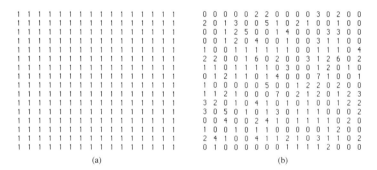

Fig. 5.13 Simulation of an Einstein solid on a microcomputer: (a) initial state; (b) thermal equilibrium state.

The number v determines the energy ε, which for a harmonic oscillator is $(v + \frac{1}{2})\hbar\omega$. One could shift the zero of energy by $\frac{1}{2}\hbar\omega$, which would yield the even simpler formula $\varepsilon = v\hbar\omega$. As for m, it is the number of oscillators all with the same energy ε; we write it as $m(\varepsilon)$, and call it the occupation number (or occupancy) of the energy state ε. One recognizes the shape of a familiar histogram (Fig. 5.6c) representing the Boltzmann distribution in the form

$$\bar{m}(\varepsilon) = \bar{m}(0) \exp\left(-\frac{v\hbar\omega}{kT}\right).$$

† In this particular case 1000 steps are amply sufficient, but we have taken 20 000 in order to verify that the thermal equilibrium is stable.

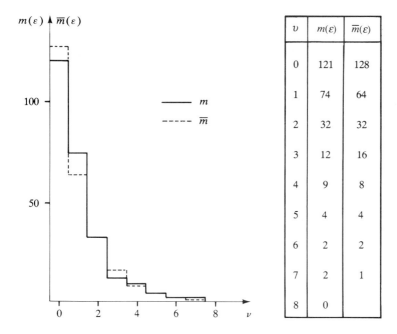

Fig. 5.14 The thermal equilibrium distribution in the simulation of the Einstein solid.

In our example the temperature corresponds to $kT = \hbar\omega/\ln 2$, whence the canonical distribution here reads simply

$$\bar{m}(\varepsilon) = \bar{m}(0) \exp(-v \ln 2) = \bar{m}(0)2^{-v}.$$

In the histogram in Fig. 5.14, the vertical coordinates are the instantaneous values $m(\varepsilon)$ given by the computer experiment (full line), and the mean values $\bar{m}(\varepsilon)$ given by the canonical distribution (broken line). The agreement is excellent; the deviations merely correspond to the expected statistical fluctuations. They could be reduced by taking more cells; but we recall that taking 9 times as many would reduce the fluctuations only by a factor of 3.

Though we have described only the principles of the simulation, it should be stressed that the model has many applications, among them the following.

1. The total number of quanta determines the future of the macroscopic system: whatever their initial distribution over the cells, one always arrives at the same thermal equilibrium state, characterized by the same temperature T.

2. The shape of the distribution is governed by the temperature; the higher the temperature, the more slowly decreasing the distribution.

3. On transferring quanta at random between two arrays at different temperatures, heat flows spontaneously from hot to cold until the temperatures equalize, through a net loss of quanta from the hotter array to the colder.

5.9 Assessment of the model

Boltzmann's statistical physics systematizes a theoretical approach that proves extremely productive, in that it leads from physics on the atomic to physics on the laboratory scale; it also leads from chance through ignorance to the determinism that governs mean values. One begins to recognize the outlines of the crucial idea that matter is structured on several different levels; and that it is the task of the physicist to discover not only the laws that operate on each level, but also the relations between different levels. It is in this respect that Boltzmann's role was so crucial. (Fig. 5.15).

There is one rather delicate point still needing discussion, namely our use of distinguishable particles in introducing the canonical distribution. This has created difficulties for certain theoretical calculations of the entropy. Since then, quantum mechanics has formulated the concept of indistinguishable particles, which leads to two new statistical distributions [14].

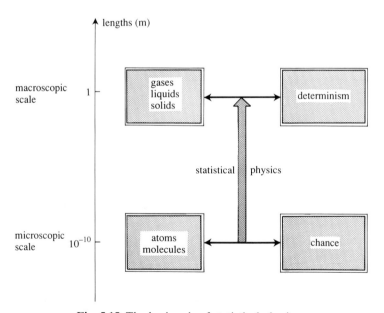

Fig. 5.15 The basic role of statistical physics.

- The Bose–Einstein distribution applies to indistinguishable particles not subject to the exclusion principle. In particular, it enables one to explain black-body radiation, and the superfluidity of liquid helium below 2.19 K.

- The Fermi–Dirac distribution applies to indistinguishable particles that do obey the exclusion principle. It enables one, for example, to study the behaviour of a gas of free electrons in the crystal lattice of a metallic conductor.

These quantum statistics are essential for describing the behaviour of matter at low temperatures. On the other hand, as the temperature rises, both tend to a common limit, the so-called 'classical limit', which corresponds to the Boltzmann distribution; hence it is the latter that one generally encounters at room temperatures, which is precisely why Boltzmann's approach has proved so fruitful.

5.10 Boltzmann-in-ordinary

As we turn these pages we increase in knowledge and therefore in our command of the laws of chance; and that in turn encourages us to take a look at the risks of everyday life, injecting an element of rationality into a domain generally considered as barred to any such thing. The point is made by the philosophical parable of the welcoming but overburdened hostess, and it concludes this chapter.

Madame Dupont likes having guests, and takes their entertainment very seriously. But her perfectionism gives her trouble, and in order to treat every guest with due attention she wants to receive only once a week, making a particular effort to arrange the dates of these occasions. First she telephones the Dubois and asks them for the last week in April, which they accept immediately. Then she telephones the Durands to ask them for the first week of May. But the Durands reply, 'Sorry, we shall be away for twelve months from the first of May. But we do very much want to see you before we leave, and should like to invite ourselves in the last week of April.' Madame Dupont sighs, inaudibly, and agrees. As soon as she replaces the receiver the phone rings: it is the Martins, to announce that they shall be passing through, and will drop in, in the last week of April. And when this much-dreaded week finally arrives, Madame Dupont has not only the Dubois, the Durands, and the Martins; she also has the Leblancs, who descend on her without so much as a phone call, because their car has broken down in the vicinity. When this marathon is over, Madame Dupont is completely exhausted, groaning that 'For two whole months no-one called, and now we have four visits one after the other, all in the same week! There is no way such a thing can have happened in the normal course of events!'

Convinced that she is the victim of some sinister machinations, she goes to consult Professor Boltzmann. The Professor listens to her attentively, and then advises her as follows. 'Madame, contrary to what you suspect, your rush of visitors is quite normal, and readily accounted for by my theory. I exclude from the year the four weeks of August when you are away on holiday; this leaves 48 weeks, in which you would like to have 48 visits, distributed uniformly at one per week. But in fact the visits cannot be prearranged, because their timing depends on many different factors operating independently; let us therefore assume that they are distributed at random. If they are, then I notice at once that the uniform distribution, corresponding to just one visit every week, is the least probable of the lot. The most probable distribution is the canonical one, which I derived in 1872 and which is named after me. For next year it predicts, distributed randomly over the year,

24 weeks with no callers, for you to relax in;
12 weeks with one caller, which is the average rate;
6 weeks with two callers, which is already a bit of a strain;
3 weeks with three callers, which will take it out of you even more;
2 weeks with four callers, which will grind you down completely;
1 week with five callers, whose effects I dare not even imagine.

As you see, dear lady, statistical physics provides a rational explanation of your experience.'

Whereupon Madame Dupont goes home to face her visitors to come, freed from her apprehensions on the metaphysical plane but very distressed by scientific prediction, and firmly convinced that all her troubles are the fault of Professor Boltzmann.†

---

† One is reminded of the simulation of the Einstein solid from the preceding chapter. Each week is represented by one cell and each caller by one quantum. For 48 cells and 48 quanta the predicted distribution reads

$$\bar{m}(v) = \bar{m}(0) \exp(-v \ln 2) = \bar{m}(0) 2^{-v},$$

where $m(v)$ is the number of weeks with v callers. This yields the sequence $\bar{m}(v)$ of mathematically expected values: 24; 12; 6; 3; 1.5; 0.75; . . . . In the parable we have for simplicity upped 1.5 to 2 and 0.75 to 1.

Appendices to Chapter 5

1 Irreversible adiabatic expansion with 20 molecules

We treat the problem like a game of heads or tails:

number of molecules in region A: m_A
number of molecules in region B: m_B $\Big\}$ $m_A + m_B = 20$;

total number of macro-states: 21;
total number of micro-states: $2^{20} = 1\,048\,576$;
total number of micro-states corresponding to the macro-state (m_A, m_B):

$$\binom{n}{m_A} = \frac{n!}{m_A!\, m_B!} = \binom{20}{m_A};$$

probability assigned to region A: $a = 0.5$;
probability assigned to region B: $b = 0.5$;
probability assigned to any one micro-state: $a^{m_A} b^{m_B} = 2^{-20}$;
probability assigned to the macro-state m_A:

$$P(m_A) = \binom{n}{m_A} a^{m_A} b^{m_B} = \binom{20}{m_A} 2^{-20};$$

average value of m_A: $\bar{m}_A = na = 10$;
standard deviation: $\sigma = (nab)^{1/2} = 5^{1/2} = 2.24$.

All these results are listed in the table below.

m_A	$\binom{20}{m_A}$	$P(m_A)$	m_A	$\binom{20}{m_A}$	$P(m_A)$	m_A	$\binom{20}{m_A}$	$P(m_A)$
0	1	0.000 001	7	77 520	0.073 929	14	38 760	0.036 964
1	20	0.000 019	8	125 970	0.120 134	15	15 504	0.014 786
2	190	0.000 181	9	167 960	0.160 179	16	4 845	0.004 621
3	1 140	0.001 087	10	184 756	0.176 197	17	1 140	0.001 087
4	4 845	0.004 621	11	167 960	0.160 179	18	190	0.000 181
5	15 504	0.014 786	12	125 970	0.120 134	19	20	0.000 019
6	38 760	0.036 964	13	77 520	0.073 929	20	1	0.000 001

2 Stirling's formula

The factorial of m is defined as

$$m! = 1 \times 2 \times 3 \times \cdots \times (m-1) \times m.$$

Stirling's formula yields an easy-to-use approximation to the natural logarithm ln $m!$ of $m!$. In order to derive the approximation we consider the graph of ln x as a function of x (Fig. 5.16). All along the curve we have drawn a sequence of rectangles having width 1 and heights ln 2, ln 3, ln 4, etc. The combined area of these rectangles, shown shaded, is

$$\ln m! = \ln 1 + \ln 2 + \ln 3 + \cdots + \ln(m-1) + \ln m.$$

When m is very large, it is an excellent approximation to replace the shaded region by the area between the x-axis and the curve ln x, which is

$$\int_1^m \ln x \, dx.$$

Integration by parts yields

$$\int_1^m \ln x \, dx = \left[x \ln x \right]_1^m - \int_1^m x \frac{dx}{x},$$

whence

$$\int_1^m \ln x \, dx = m \ln m - m + 1.$$

Since m is very large we can drop the 1, and we are left with

$$\ln m! \approx m \ln m - m.$$

These are the first two terms in Stirling's formula. A somewhat more elaborate argument yields a third term, giving

$$\ln m! \approx m \ln m - m + \tfrac{1}{2} \ln 2\pi m.$$

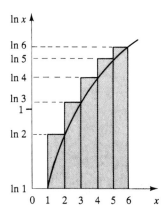

Fig. 5.16 Simple illustration of Stirling's formula (the vertical coordinates have been multiplied by 5 in order to make the pattern easier to see).

However, very often two terms are quite enough, say when m is of the order of Avogadro's number and therefore very large.

Finally we note that differentiation of the formula

$$\ln m! = m \ln m - m$$

yields a result already derived in Appendix 4 of Chapter 2:

$$\frac{d(\ln m!)}{dm} = \ln m.$$

3 The Boltzmann distribution

The method we describe is Boltzmann's own. Later, a more powerful method was devised by Gibbs [14].

Our object is to determine the set of occupation numbers m_i which maximizes the number Ω of accessible micro-states. The governing equations are

$$\Omega = \frac{n!}{\prod_i m_i!}, \qquad \sum_i m_i = n_i, \qquad \sum_i \varepsilon_i m_i = U.$$

We look for the maximum of $\ln \Omega$ by exploiting

$$\ln m! = m \ln m = m,$$

derived in the preceding section. One finds

$$\ln \Omega = \ln n! - \sum_i \ln m_i!,$$

whence

$$\ln \Omega = n \ln n - n - \sum_i m_i \ln m_i + \sum_i m_i.$$

Next, in view of $\sum_i m_i = n$, we have

$$\ln \Omega = n \ln n - \sum_i m_i \ln m_i.$$

Differentiation yields

$$d(\ln \Omega) = 0 - \sum_i m_i \frac{dm_i}{m_i} - \sum_i \ln m_i \, dm_i.$$

The second term $\sum_i dm_i$ is just the differential $d\sum_i, m_i = dn$, which vanishes because n is constant. We are left with

$$d(\ln \Omega) = -\sum_i \ln m_i \, dm_i.$$

If Ω is a maximum then $d(\ln \Omega)$ must vanish, which yields a first condition, namely

$$\sum_i \ln m_i \, dm_i = 0.$$

The dm_i must satisfy two further conditions:

$$\sum_i m_i = n = \text{constant, whose derivative yields } \sum_i dm_i = 0;$$

$$\sum_i m_i \varepsilon_i = U = \text{constant, whose derivative yields } \sum_i \varepsilon_i \, dm_i = 0.$$

Accordingly, the following three conditions must be satisfied identically (i.e. for arbitrary values of the dm_i):

$$\sum_i \ln m_i \, dm_i = 0, \qquad \sum_i dm_i = 0, \qquad \sum_i \varepsilon_i \, dm_i = 0.$$

On introducing two Lagrange multipliers, provisionally called α and β, these three conditions can be amalgamated into the single condition

$$\sum_i (\ln m_i + \alpha + \beta \varepsilon_i) \, dm_i = 0.$$

This can hold for arbitrary dm_i only if the contents of every pair of brackets vanish, which implies that each variable m_i takes a value m_i' satisfying the relation

$$\ln m_i' + \alpha + \beta \varepsilon_i = 0,$$

or in other words

$$\ln m_i' = -\alpha - \beta \varepsilon_i.$$

It is these values m_i' that describe statistical equilibrium.

One exponentiates by writing $A = \exp(-\alpha)$, and finds

$$m_i' = A \exp(-\beta \varepsilon_i).$$

This is the distribution law for the most probable m_i, which warrants it as the equilibrium distribution. It remains only to determine A and β.

- A is easily found by writing

$$n = \sum_i m_i' = \sum_i A \exp(-\beta \varepsilon_i) = A \sum_i \exp(-\beta \varepsilon_i).$$

The combination $\sum_i \exp(-\beta \varepsilon_i)$ is called the partition function, and denoted by Z; it follows that $A = n/Z$.

- As regards β, extended analysis eventually leads to $\beta = 1/kT$, where k is

Boltzmann's constant and T the thermodynamic temperature, which is a macroscopic variable measured with a thermometer [14].

Accordingly, our conclusions in full read

$$m'_i = \frac{n}{Z} \exp\left(-\frac{\varepsilon_i}{kT}\right).$$

Finally, on switching from occupation number m'_i to occupation probability $P(\varepsilon_i)$ through $P(\varepsilon_i) = m'_i/n$, one obtains

$$P(\varepsilon_i) = \frac{\exp(-\varepsilon_i/kT)}{Z}.$$

This is the Boltzmann distribution.

Further reading

See Adler and Wainwright (1959), Guinier and Jullien (1989), Kittel (1969), Messiah (1961–62), and Rocard (1961) in the Bibliography.

Henri Poincaré. French mathematician; born in Nancy, died in Paris (1854–1912). He makes important contributions to all the branches of the mathematics of his day: non-Euclidean spaces, automorphic functions, differential equations, integral equations, calculus of probabilities, analytic mechanics. In addition to all this, he thinks deeply about philosophy; his books and articles are still essential reading in epistemology: *La science et l'hypothèse*; *Science et méthode*; *La valeur de la science.* An important chapter of his mathematical discoveries is published in 1892 under the title *Méthodes nouvelles de la mécanique céleste* (New methods of celestial mechanics). It contains in particular his method for displaying the properties of solutions of nonlinear differential equations. He shows how such equations lead one to systems very sensitive to initial conditions, anticipating that models of this kind will one day be important in meteorology. (Boyer-Viollet)

6

Poincaré, or deterministic chaos (sensitivity to initial conditions)

Chaos umpire sits,
And by decision more embroils the fray
By which he reigns

Milton

In the preceding chapter we saw, with Boltzmann, how chance could generate determinism. The present chapter will show how determinism can generate chance. This domain was pioneered by Henri Poincaré near the end of the last century, but it is only the advent of fast computers in the early 1960s that has made its development possible. Today it is a very active field of research, and has induced a far-reaching revolution in our concepts. In order to elucidate it, we start by making a semantic distinction between random and chaotic processes respectively.

- A dust particle suspended in water moves around randomly, executing what is called Brownian motion. This stems from molecular agitation, through the impacts of water molecules on the dust particle. Every molecule is a direct or indirect cause of the motion, and we can say that the Brownian motion of the dust particle is governed by very many variables. In such cases one speaks of a *random process*; to treat it mathematically we use the calculus of probabilities described in Chapter 2.

- A compass needle acted on simultaneously by a fixed and by a rotating field constitutes a very simple physical system depending on only three variables. However, we shall see that one can choose experimental conditions under which the motion of the magnetized needle is so unsystematic that prediction seems totally impossible. In such very simple cases whose evolution is nevertheless unpredictable one speaks of *chaos* and of *chaotic processes*; these are the terms we use whenever the variables characterizing the system are few.

6.1 Sensitivity to initial conditions

The gunner's problem was described in Chapter 1, and enabled us to define what we mean by determinism (Fig. 6.1).

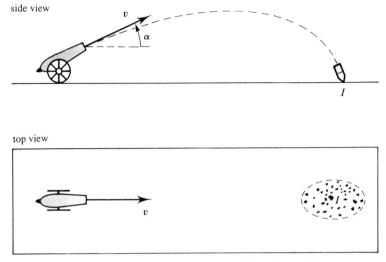

Fig. 6.1 The evolution of gunnery leads to determinism.

On firing the gun many times, while trying to keep the initial conditions (position and velocity) as constant as possible, one observes that the impacts are spread over a scatter zone, an ellipse centred on I. The story of gunnery is the story of successive technological advances enabling one to determine the initial conditions more and more precisely, in order to reduce the scatter zone. Technical improvement stops when the ellipse is no larger than the target, at which point the gunner considers his problem as solved. Physicists and philosophers however are more demanding; they imagine further improvements, determining the initial conditions ever more closely, with correspondingly diminishing deviations of the impacts from the point I. Finally they envisage an idealized limit corresponding to absolute precision: absolutely constant initial conditions, with every shell landing exactly on the point I. It is with this ideal limit that we associate the notion of determinism, that is of a method allowing one to predict the future exactly, from initial conditions that are likewise exact. The best example of a deterministic theory is classical mechanics. But this definition of determinism is based on the tacit assumption that the deviations on arrival diminish roughly in proportion to the deviations at departure. In that case the idealized limit can be envisaged quite clearly; and in fact gunners can realize excellent approximations to it. But what would happen if initial and final deviations were connected by a relation more complex than simple proportionality? How much would then survive of determinism, and of our ability to predict the future of the system? One can understand this better from a concrete example.

Consider a variant of billiards, with seven consecutively numbered cylindrical stops fixed rigidly to the table in the staggered configuration shown in Fig. 6.2. The ball is placed at the point having coordinates (x, y), and is then projected at some angle α to the direction Ox. The jackpot is won by the player who, in a single shot, can cause the ball to touch all seven stops in their proper order.

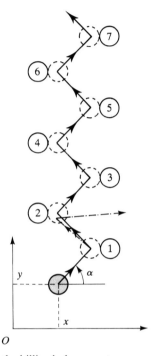

Fig. 6.2 A difficult task for the billiard player: extreme sensitivity to the angle α of the shot makes the task all but impossible.

The prospect of a jackpot is good reason for serious reflection about our capacity to predict the future of the system. Two questions come to mind immediately.

(a) Is there a solution at all?

(b) Is this a game of skill or a game of chance?

To answer the first question we use a simplified model based on the following assumptions.

1. The ball rolls without slipping, that is without losing any kinetic energy,

and is given no other spin; in that case its centre describes a straight line between any two successive collisions.

2. The collisions with the fixed stops are elastic, whence the angle of reflection equals the angle of incidence.

Though this rather elementary model is only an approximation to the truth, it does suggest that the answer to our first question is yes: the aim is achievable provided we can make the ball describe the ideally regular saw-tooth trajectory shown in Fig. 6.2.

In trying to answer the second question, nothing at first promises better than an experimental approach, whose results nevertheless are deceptive! Average players generally succeed in hitting stops 1 and 2, after which the ball strays (Fig. 6.2). A player of international standard, commanding great precision of movement, may hit stops 1 to 4. Nobody, however, can do better, for skill has its limits. Is it possible that a mechanical launcher could reach 5 or 6? Be that as it may, precision will fail before stop 7, and any confidence that we can hit the jackpot vanishes. We could, of course, persevere, trying with every shot to approximate the ideal trajectory as closely as possible. In this way we might even succeed. But even if we did, we should ascribe the win to chance, because there would have been nothing before the final shot that would have allowed us to predict its success. What attitude should one adopt in this situation?

The key to the problem lies in the shapes, spherical for the ball and cylindrical for the stops: though success can indeed be achieved by an ideal trajectory determined by the initial conditions x, y, and α, any deviation from these exact initial conditions, however small, produces a trajectory that is very different from the ideal. Thus, an initial deviation ε from the angle α is multiplied by 9 at each reflection, which very quickly leads to failure.† For instance, with $\varepsilon = 0.5°$, the player fails by a wide margin after stop 2 (Fig. 6.2). The international player would need to command a precision of $\varepsilon = 2.6'$ in order to be sure of reaching stop 4.‡ Finally, if one wanted to be sure of reaching stop 7, one would need to attain the quite inconceivable precision $\varepsilon < 0.22''$.

In order to hit the jackpot all the same, our only course is to embark on a

---

† We reach this conclusion by exploiting a result characteristic of spherical and of cylindrical convex mirrors:

$$\varepsilon' = \varepsilon \left(1 + \frac{2d}{R \cos \phi} \right);$$

where ϕ is the angle of incidence of the pencil of rays, ε is the opening angle of the pencil before reflection, ε' is the opening angle of the pencil after reflection, d is the distance from light-source to mirror, and R is the radius of curvature of the mirror.

‡ The centre of the ball should touch a circle of radius $2R$ centred on stop 4. At the point of impact on stop 3, this circle subtends an angle of $32''$. The allowable initial deviation ε is obtained on dividing by 9^3.

series of trials, continued until we achieve the success which is possible but purely fortuitous; here therefore we speak of chance by reason of our inability to realize the requisite initial conditions with sufficient precision. Only absolute precision at the start would suffice to predict the future of the system with any exactitude. Because of its exponentially rapid growth in time, the slightest initial deviation makes any long-term prediction impossible.

Accordingly, we are faced with a macroscopic system, modelled by a perfectly deterministic rule (specular reflection), which is nevertheless so sensitive to the precise initial conditions that in practice its long-term future is unpredictable. And yet the system depends on only three variables x, y, and α (Fig. 6.2). Thus, the system we are dealing with is chaotic, and such systems are *unpredictable*.

When faced with random phenomena, physicists, far from retiring defeated, proceeded to devise the calculus of probabilities in order to formulate the laws characterizing such processes. For chaotic processes, our procedure is similar, aiming to disentangle the qualitative and the quantitative rules that govern this novel domain.

6.2 The conceptual framework

The example of the ordinary pendulum (Fig. 6.3) serves to introduce the modest vocabulary needed by the compleat physicist in order to discuss chaotic processes.

A *dynamical system* is a physical system that moves, as the pendulum obviously does.

The *degrees of freedom* are the coordinates required to describe the exact position of the system in space. The pendulum for instance has one degree of freedom, namely the angular displacement θ. The position variables alone are not enough to define the dynamical state of the system: each position variable must be supplemented by its time-derivative, which is a different kind of variable, namely a speed. Accordingly, the dynamical state of the pendulum is specified by the variables θ and $\dot{\theta}$ (Fig. 6.3). (Henceforth we call 'the dynamical state' simply 'the state' when there is no danger of confusion.)

Phase space is an abstract space whose coordinates are the position and the velocity variables of the system. For instance, the state of the pendulum at a given time is specified by a point M having the coordinates θ and $\dot{\theta}$ in a two-dimensional phase space. In the course of time the point M describes a trajectory in this space; the trajectory contains information necessary and sufficient to describe the motion of the pendulum at any instant (Fig. 6.3a). By an extension (and an abuse) of language, physicists have fallen into the habit of

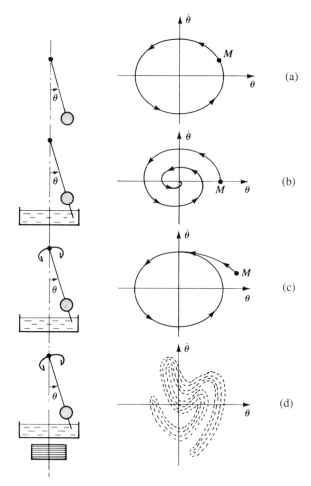

Fig. 6.3 Oscillations of a pendulum and the associated phase space: (a) frictionless pendulum → periodic trajectory; (b) damped pendulum → attractive fixed point; (c) damped and driven pendulum → strange attractor.

saying that the pendulum is 'a system with two degrees of freedom in phase space'.

A *conservative system* is a dynamical system without energy dissipation (i.e. not subject to damping); its mechanical energy is constant in time. A heavy pendulum, carefully suspended from a knife edge and swinging in a vacuum, oscillates very regularly for a very long time, because it experiences very little friction. It is a good example of a conservative system, its phase space trajectory being a closed ellipse repeated once every period (Fig. 6.3a).

A *dissipative system* is a dynamical system subject to damping; its mechanical energy is gradually dissipated as heat. To illustrate this notion, the bob of the pendulum is provided with a paddle which dips into an oil bath, where it experiences viscous resistance. Then the oscillations of the pendulum decreases in amplitude over each successive swing, until it reaches a state of complete rest with $\theta = 0 = \dot{\theta}$. The phase space trajectory is a spiral, winding progressively inward towards the origin O: the process evolves as if the point M were attracted by the point O (Fig. 6.3b).

A *driven dissipative system* is a dynamical system whose mechanical energy loss due to dissipation is compensated by mechanical energy supplied from outside. The escapement mechanisms invented by clockmakers allow the pendulum to continue to swing: once it is launched with enough initial momentum, its motion very soon settles into a regime of stable oscillation. In phase space the point M is seen to move rapidly towards a limit cycle: the system evolves as if M were attracted by the limit cycle (Fig. 6.3c).

An *attractor* is a region of phase space where the representative point M eventually ends up irrespective of the initial conditions.† Only dissipative systems have attractors, and we have met two kinds already: fixed points and limit cycles. One novel class of attractor remains to be described; their geometry is so surprising that physicists have come to call them 'strange attractors'.

Figure 6.3d represents a dissipative system undergoing forced oscillation. In this example the damping is due to friction between paddle and oil bath, and the forcing to an alternating magnetic field induced by a coil, and acting on the material of the pendulum, which is ferromagnetic. One can easily find experimental conditions under which the pendulum appears to go crazy, in that its motion becomes completely irregular. Moreover, it proves impossible to repeat such an evolution twice running, however carefully one tries to reproduce the exact initial conditions. What happens in phase space? The system now has three degrees of freedom, because to the already familiar θ and $\dot{\theta}$ we must adjoin the phase ϕ of the magnetic driving field. Hence the attractor to be determined is three-dimensional, and its pictorial representation can prove very difficult. In such cases Poincaré recommends the use of cross-sections of phase space taken at constant ϕ; such *Poincaré sections* have two dimensions, θ and $\dot{\theta}$, and their appearance is very strange, presenting a multitude of points suggestive of a highly folded structure (Fig. 6.3d).

The *strange attractor*‡ is now easily defined. The notion relates to dissipative dynamic systems with only a few degrees of freedom, very sensitive to initial conditions, and behaving chaotically: for such a system, a strange attractor is a restricted region of phase space where all trajectories eventually

---

† Or at least for a very large range of initial conditions, said to constitute the 'domain of attraction' (catchment area) of the attractor.

‡ The name was first proposed in 1971 by David Ruelle and Floris Takens, in a paper that identified and elucidated the idea.

accumulate, without ever intersecting.† Strange attractors have a typical folded structure; we shall show later that this structure is a fractal (Section 6.3 and Appendices 1 and 2).

We have now outlined all the requisite new concepts and nomenclature; our object was to exorcise this vocabulary as painlessly as possible. If you feel that the exorcism has not been effective enough, you should not hesitate to reread the present section before proceeding to the next.

6.3 The crazed compass needle

Contrary to our normal habits, we have started with the semantics rather than the physics; this will however allow us a readier view both of the experimental and of the theoretical aspects of chaotic processes. Armed with our brief vocabulary we can go straight to the most typical case, the compass needle acted on simultaneously by two magnetic fields, one fixed and the other rotating. This simple example has the advantage that it affords a very spectacular demonstration experiment (see Appendix 3).

A top view of the apparatus is shown in Fig. 6.4. The magnetic needle swivels on a vertical axis, whence its motion consists of rotation in a horizontal plane, described by the angular coordinate θ. The needle is surrounded by four identical coils aligned in pairs along two mutually perpendicular axes $x'x$ and $y'y$; they generate two magnetic fields, namely

(a) a fixed field, of constant magnitude B_1 and parallel to $x'x$;

(b) a field of constant magnitude B_2 rotating uniformly with angular velocity ω_0, and specified by the angular coordinate $\phi = \omega_0 t$.

In practice one generates these two fields by passing a current

$$i_x = I_1 + I_2 \cos \omega_0 t$$

through the coils on the $x'x$ axis, and a current

$$i_y = I_2 \sin \omega_0 t$$

through those on the $y'y$ axis.

Quite simple analysis suffices to predict the behaviour of the compass needle qualitatively. We distinguish three cases.

---

† If a point of intersection did exist, then it could be taken to define initial conditions for the subsequent motion, admitting two different phase space trajectories. But this would contravene the determinism we assume in studying such systems.

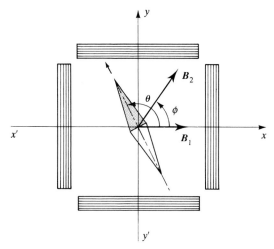

Fig. 6.4 A compass needle undergoing forced oscillations is one example of a chaotic system.

• When only the fixed field is present (B_1 non-zero, $B_2 = 0$), we have the classic situation: if the needle is slightly displaced from equilibrium and then released, it executes simple harmonic oscillations of frequency ω_1 around the direction of the fixed field $\boldsymbol{B}_1$.

• When only the rotating field is present ($B_1 = 0$, B_2 non-zero), an observer turning with the vector $\boldsymbol{B}_2$ sees exactly the same situation as we had in the preceding case: the needle oscillates around $\boldsymbol{B}_2$ with frequency ω_2. An observer fixed with respect to $x'x$ sees the motion of the needle as a combination of a rotation with angular velocity ω_0 and an oscillation with angular frequency ω_2. The overall impression is that of an irregular rotation, because, as it rotates, the needle is sometimes ahead and sometimes behind the field $\boldsymbol{B}_2$; only on average is the rotation uniform.

• When both fields are present (B_1 and B_2 both non-zero), and if the angular velocity ω_0 is not close to either of the resonance values† ω_1 and ω_2, then the motion becomes more complicated. The position of the needle is very

† It can be shown that

$$\omega_1 = \left(\frac{\mathcal{M} B_1}{J}\right)^{1/2} \quad \text{and} \quad \omega_2 = \left(\frac{\mathcal{M} B_2}{J}\right)^{1/2},$$

where $\mathcal{M}$ is the magnetic moment of the compass needle, and J is the moment of inertia of the needle around its (vertical) axis of rotation.

unstable, because it is coupled preferentially sometime to B_1 and sometime to B_2. The resulting motion appears to be completely disorganized. Moreover, because the system is so unstable, and because it is impossible in practice to release the needle twice in exactly the same way (same θ, same $\dot{\theta}$, same ϕ), one can never repeat exactly the same sequence of this disorganized motion twice running. Therefore we are faced with a system very sensitive to its initial conditions even though it depends on only three variables, and its motion is chaotic.

Figure 6.5 shows a side view of the apparatus. The axle of the compass needle is mounted on two ruby bearings, with the lower bearing immersed in oil. The viscosity of the oil is adjusted by adjusting its temperature, which allows the damping of the needle to be controlled very accurately. Thus one can go continuously from an almost undamped (i.e. practically conservative) to a heavily damped system with strong dissipation. Meanwhile, the system is driven, because its coupling to the rotating field allows it to replenish the energy lost through damping. The four coils that generate the fields B_1 and B_2 are not shown. Note, however, the presence of two further coils connected in series; these act as detectors of the needle's motion, by delivering an induced current proportional to the angular velocity $\dot{\theta}(t)$. The oscillator with frequency ω_0 defines the time-base. It has three ports; the first feeds the coils that generate the rotating field; the second differentiates the signal $\dot{\theta}(t)$; and the third generates a signal having the phase $\phi = \omega_0 t$. The three signals $\dot{\theta}(t)$, $\ddot{\theta}(t)$, and $\phi(t)$ are fed into a microcomputer† programmed to furnish, on demand, a Poincaré section in the plane of the variables $\dot{\theta}(t)$ and $\ddot{\theta}(t)$; the output is observed on a screen.‡ The result, shown in Fig. 6.6, is a very pretty example of a strange attractor; and it is no mean experimental feat (hinging on very careful suspension of the needle) to have succeeded in demonstrating this very characteristic folded structure so clearly [4].

By virtue of the theorem about the rate of change of angular momentum, the equation of motion of the system is very simple:

$$J\frac{d^2\theta}{dt^2} + f\frac{d\theta}{dt} = -\mathcal{M}B_1 \sin\theta - \mathcal{M}B_2 \sin(\theta - \omega_0 t),$$

where J is the moment of inertia of the compass needle, f is a friction constant, and $\mathcal{M}$ is the magnetic moment of the needle.

The terms $-\mathcal{M}B_1 \sin\theta$ and $-\mathcal{M}B_2 \sin(\theta - \omega_0 t)$ represent the couplings to the fixed and to the rotating fields repectively. The differential equation is

---

† After analog to digital conversion, which transforms the original amplitude into a binary-coded signal.

‡ The phase space $(\theta, \dot{\theta}, \phi)$ is more familiar, but for practical reasons the experimenter here has chosen the space $(\dot{\theta}, \ddot{\theta}, \phi)$. The choice is perfectly convenient, since it displays the strange attractor equally well.

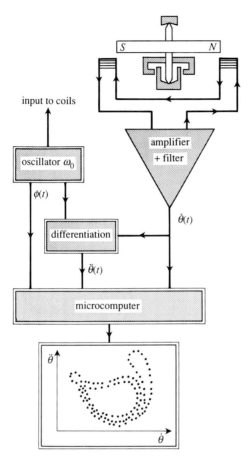

Fig. 6.5 Sketch of the experimental layout for constructing the strange attractor of a damped compass needle undergoing forced oscillations. The coils responsible for the fields B_1 and B_2 are not shown.

nonlinear; it has no analytic solution, but can be integrated numerically. Such integration yields the strange attractor and its Poincaré sections, with far more precision than the direct experimental method just described. It enables one to see that the folded construction is repetitive, being self-similar whatever the scale on which it is observed; it follows that the structure of the strange attractor is that of a fractal (see Appendices 1 and 2).

We are now in a position to spell out the difference in behaviour between a random and a chaotic system.

• The trajectory of a random system is spread uniformly through all the accessible regions of phase space, with no accumulation in any particular region.

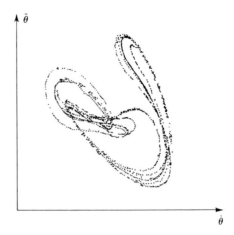

Fig. 6.6 The strange attractor of a damped compass needle undergoing forced oscillations (after Vincent Croquette).

• The trajectory of a chaotic system contracts very quickly on to a restricted region, namely on to the strange attractor; accordingly there is a certain order in such chaos.

With the help of the damped compass needle undergoing forced oscillations, we have displayed all the characteristics of chaotic systems. The system is simple, having been devised with that end in view, and it has served its purpose perfectly. Now of course we must ask whether chaotic systems operating on the same principles really occur in nature (and not only as artefacts made by physicists). This question leads us to study turbulence.

6.4 Turbulence

The way taps work is most instructive; anyone who has never noticed this should transfer to the kitchen until the truth of the observation becomes evident.

• When the tap is barely open, the water emerges in the form of a very regular thread: this is laminar flow.
• When the tap is fully open, the jet twists and is very irregular: this is turbulent flow.

On the theory of laminar flow we have a good grip; by contrast, turbulence remains one of the most challenging problems in theoretical physics. Though the basic equations of fluid dynamics are well known, every practical application leads to a system of differential equations that cannot be solved, at least not exactly. At present, the trend is to try and progress towards an understanding of turbulence through deep analysis of the various scenarios for the transition from laminar to turbulent flow. Here we present one such scenario, chosen because it is simple and quite typical.

Though meteorology is a branch of fluid dynamics whose importance to mankind is unarguable, it is common knowledge that weather-forecasting is far from perfect. One can see this from the regional forecasts shown every evening on television. On the one hand, it is announced that tomorrow the sun will rise at 6.47 hours, which is an astronomical prediction, precise and trustworthy. On the other hand, it is also announced that there will be scattered showers throughout the region, without indicating either time or place: the contrast is striking. The (French) national meteorological service broadcasts more detailed bulletins for farmers, for aviation, and for shipping. To this end it has installed a supercomputer, a CRAY II, which allows more precise prediction, extending up to five days ahead. However, one gets the impression that exponential increases in technical support are required to secure merely linear increases in the forecasting period; and the mind boggles if one tries to contemplate computers large enough to predict eight days ahead with reasonable confidence. In other words, weather forecasting is possible only in the short term; in the medium and long term it fails, in spite of the fact that meteorologists work just as conscientiously as astronomers. Why, therefore, such flagrant disparity?

This question is answered very clearly by the model of turbulent convection proposed by Edward Lorenz in 1963. Lorenz chose an elementary model of the Earth's atmosphere, which nevertheless affords rather general conclusions. He envisages a large plain covered by uniform vegetation.† Sunlight heats the ground, establishing a temperature gradient between zero height, where it is hot, and the upper atmosphere, which is cold. Lorenz aims to write down equations from which to calculate the motions of the atmosphere over this plain as governed by the temperature gradient. To this end he simplifies the equations of fluid dynamics as much as possible, and is eventually led to describe the motions with the aid of only three suitably normalized variables:

$x(t)$, the amplitude of the convective motion;

$y(t)$, the temperature difference between ascending and descending air currents;

$z(t)$, the difference between the temperature given by the model and that which would follow if temperatures varied linearly with height.

---

† A plain excludes perturbations due to varying ground relief. Uniform vegetation excludes differential heating due to differences in the absorption coefficient for sunlight.

In each case, t stands for time.

In terms of these variables, Lorenz's equations read

$$\frac{dx}{dt} = \sigma(y-x), \qquad \frac{dy}{dt} = rx - y - xz, \qquad \frac{dz}{dt} = -bz + xy,$$

where σ, b, and r are parameters. Lorenz chose $\sigma = 10$ and $b = \frac{8}{3}$; then he looked at how solutions evolve for different values of r, the parameter that specifies the temperature gradient. We are faced with a system of (coupled) differential equations. Without the terms xz and xy, all three equations would be linear; such a system would be integrable, and one could find an analytic solution for x, y, and z. But the actual equations are not linear, and solutions can be found only by numerical integration on a computer. That is how Lorenz found three different kinds of solution, which can be represented graphically in a fictitious three-dimensional space with coordinates x, y, and z.†

• For $0 < r < 1$, the temperature gradient is weak, corresponding to a calm night. The air is still, and small perturbations such as might be produced by a moving animal are quickly damped out. What we see in our fictitious space is the behaviour of a dissipative system evolving towards an attractor, namely towards the fixed point O. This point indeed corresponds to still air: there is no convection ($x=0$); no temperature differences between ascending and descending air currents, since there are no currents ($y=0$); and the temperature varies linearly with height ($z=0$) (see Fig. 6.7a).

• For $1 < r < 24$, the temperature gradient is modest. Air heated by the ground tends to rise towards the colder layers, while the colder layers naturally tend to fall; a regular and stable convective regime is established spontaneously, known as Rayleigh–Bénard convection.‡ Here this regime takes the form of convective rolls with their axes horizontal, the sense of circulation alternating between adjacent rolls. This situation corresponds to a regular pattern of winds at ground level and above, and to the presence of the rising and falling air currents well known to gliders. In our fictitious space there are two fixed-point attractors, one corresponding to circulation in one and the other to circulation in the opposite sense; every phase-space trajectory necessarily runs into one or other of these two points (Fig. 6.7b).

• For $24 < r < 148$, the temperature gradient is steep; the atmosphere is a stage for the disorganized motions that we call turbulence. Since the system depends on only three variables, we speak of this process as chaotic. It is very instructive to observe what happens in our fictitious space: every trajectory

---

† One must not confuse these fictitious x, y, z with the coordinates in ordinary configuration space.

‡ It was observed experimentally by M. Benard in 1900, and interpreted theoretically by Lord Rayleigh in 1916.

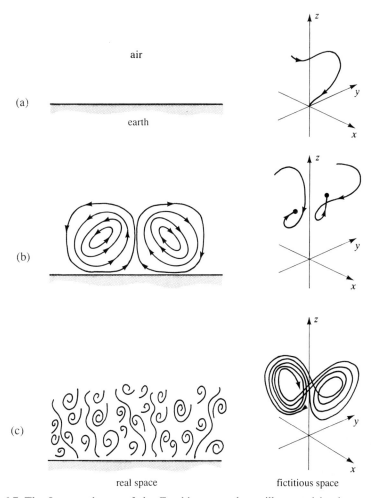

Fig. 6.7 The Lorenz theory of the Earth's atmosphere, illustrated in three typical regimes: (a) atmosphere at rest; (b) atmosphere in ordered convection; (c) atmosphere in turbulent convection.

twists into a nearly plane double spiral, having fractal dimension 2.06, this fractal structure being a characteristic of the strange attractor (Fig. 6.7c).†

Once we recognize the presence of a strange attractor we know that the atmosphere is a dissipative system very sensitive to the initial conditions. Two

† Note that Fig. 6.7c is not a two-dimensional Poincaré section, but a representation of our fictitious three-dimensional space in perspective. It can be shown that the fractal dimension is less than the number of degrees of freedom of the system. In the present case, the fractal dimension 2.06 implies at least three degrees of freedom, conformably with the facts. (Of course, these are degrees of freedom ordinary space.)

phase-space trajectories issuing, respectively, from two points very close together will diverge exponentially, and each will very quickly spread all over the attractor; but they cannot get further away from each other than this. Therefore, after only a short time interval the two trajectories become indistinguishable; the system has lost all memory of the initial conditions, and all we can assert is that the point representing it remains always confined to the strange attractor. We have now learnt enough to summarize Lorenz's conclusions.†

The motions of the atmosphere are extremely sensitive to the initial conditions; a butterfly flaps its wings in the West Indies, which provokes a storm two weeks later in the English Channel. This tongue-in-cheek but very vivid example, due to the American meteorologists Charles Leith and Robert Kraichman, is known as the 'butterfly effect'.

Precise long-term weather-forecasting is impossible, because the two-week time barrier cannot be surmounted; in order to do so, one would need unlimited precision both in the initial conditions and in the computer integrations, and both requirements are absurd.

Long-term climatological prediction on the other hand is possible, because the existence of a strange attractor shows that only certain kinds of turbulent motion will occur. The tool for long-term prediction is the calculation of physical quantities averaged by integration over the entire attractor. Since everyone is allowed to dream, we can imagine a day when we shall know the strange attractor say of the département du Rhône, and its deterministic evolution in time; that would enable us to predict, for May next year, 5 cm of precipitation, 17 sunny days, and an average temperature of 15°C. This kind of prediction would prove extremely useful for all outdoor economic activities like farming, building, transport, and so on, which underlines the value of abstract research on strange attractors, since it could lead to the solution of very concrete problems.

It remains only to answer the question asked at the start of this section: why does the astronomer generally predict so well, and the meteorologist so badly?

The astronomer studies the motion of the system Sun plus Earth, which is governed by classical determinism, at least on the scale relevant here. This means that two phase-space trajectories issuing from two neighbouring points diverge linearly with time; hence any gain in precision in determining the hour of sunrise on day D entails the same relative gain in precision for predictions for days $D+1, D+2, \ldots, D+365, \ldots$

By contrast, the meteorologist studies the motions of the Earth's atmosphere, which are governed by deterministic chaos. This means that two

† One can continue the exploration to values of r greater than 148, identifying a periodic regime ($148 < r < 166$), an intermittently chaotic regime ($166 < r < 233$), and then another periodic regime ($r > 233$).

phase space trajectories issuing from two neighbouring points diverge exponentially with time, whence any gains in precision in determining initial conditions are profitable only for short-term prediction, but improve long-term prediction not at all. However, by way of compensation, meteorologists have discovered strange attractors, which might lead them to great advances in climatology.

6.5 Assessment of the model

Even though the field is developing rapidly, a provisional assessment is quite possible.

The departure point for the discovery of deterministic chaos is the effort mathematicians have devoted to an irritating problem of celestial mechanics, namely to the *three-body problem*.† To begin with, Henri Poincaré proved that the problem has no analytic solution that is both exact and general. He followed this up by proposing semi-quantitative methods of solution: these are the famous 'méthodes nouvelles de la mécanique céleste' (new methods of celestial mechanics). More recently, Michel Hénon demonstrated that computer integration enables one to identify different kinds of solutions. These are the researches that first recognized extreme sensitivity to initial conditions, albeit they were restricted to conservative systems (there is no friction in space).

The physics community was finally convinced that deterministic chaos exists when it became apparent that such sensitivity to initial conditions also occurs in dissipative systems, that is nearer to our immediate experience, and that strange attractors occur not only in numerical simulations but also in the results of real physical experiments. For the crazed compass needle and for turbulent convection this conclusion is now well established; and it is being looked for in chemical kinetics, where systems governed by nonlinear differential equations are likewise familiar. The motion of all these systems is chaotic in real space, but transcribed into phase space it displays some semi-quantitative order, embodied in the strange attractor. Though one cannot tell where the representative point of the system will be at a given time, one can at least assert that it will be somewhere on the attractor; and it is by virtue of the attractor that prediction becomes possible (Fig. 6.8).

Philosophically speaking, our conception of chance has been enriched: in addition to chance by reason of ignorance (Maxwell and Boltzmann), we now have chance by reason of inaptitude (Poincaré), both based on determinism.

---

† The problem is to find the general analytic solution of the equations describing the motion of three celestial bodies (planets or stars) that attract each other according to Newton's law of gravitation.

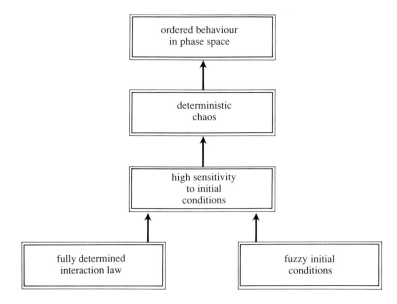

Fig. 6.8 Flow diagram for deterministic chaos.

Finally, artist and designer will certainly be charmed by strange attractors, all the more so if they watch them under construction by the computer: few processes can be as fascinating to observe as the crazy dance of the graph plotter, registering point after point on a Poincaré section with no apparent logic whatever, yet producing eventually the overall pattern of an attractor with its indisputable (if strange) beauty.

Appendices to Chapter 6

1 Fractals

A fractal is a geometric structure that repeats itself, being self-similar irrespective of the scale on which it is observed. Though the word 'similar' is used here in its precise mathematical acceptation, we shall distinguish between rigorous similarity as for the Cantor set (Fig. 6.9), and statistical similarity as for the coastline of Brittany (Fig. 6.10). These two examples will serve to flesh out the definition given above.

The *Cantor set* is the simplest example of a fractal. It is constructed by subdividing a line segment into three equal parts, and then keeping only the two flanking parts. This operation is repeated indefinitely, and the Cantor set

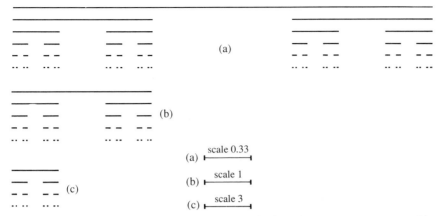

Fig. 6.9 The Cantor set is the simplest example of a fractal structure. Here parts of it are shown on three different scales.

consists of all the eventually remaining parts. Figure 6.9 shows it on three different scales, and the perfect similarity between these is evident.

The coastline of Brittany, observed vertically from a high altitude above the Crozon peninsula, presents an uninterrupted and very broken sequence of bays and headlands. Viewed from successively lower altitudes, the outline varies, but its fragmented appearance persists, with smaller headlands and smaller bays; four such changes of scale demonstrate the persistence of the structure. In this case the similarity is of a statistical kind.

2 Fractal dimension

We start by introducing the idea of dimensionality in an elementary way through lengths, areas, and volumes, that is through magnitudes whose geometry is familiar (Fig. 6.11). To this end we look for the appropriate scaling law, or in other words for the rule telling us how these magnitudes vary when all linear dimensions are multiplied by a given coefficient. To keep the discussion simple and definite, we choose this coefficient as 2.

- As regards lengths, one obtains a single segment having twice the original length, which can be constructed from two segments each equal to the original one. We say that the number of pieces required is $2 = 2^1$.

- As regards areas, starting with a square one obtains a new square having four times the original area, which can be constructed from four equal squares. We say that the number of pieces required is $4 = 2^2$.

- As regards volumes, one obtains a cube eight times larger than the original one, which can be constructed from eight equal cubes. We say that the number of pieces required is $8 = 2^3$.

scale ├──── 40 km ────┤   scale ├──── 10 km ────┤

scale ├──── 4 km ────┤   scale ├──── 1 km ────┤

Fig. 6.10 The coastline of Brittany, very broken on any scale, is an example of the statistical self-similarity that characterizes fractal geometry.

We are now in a position to identify the general rule, using the following notation: n is the number of pieces, s is the 'similarity ratio' (or 'scaling factor'), and d is the dimensionality (i.e. the number of dimensions). Then the rule reads

$$n = s^d \quad \text{or} \quad d = \frac{\ln n}{\ln s}.$$

It can be used to define the dimensionality, and we see at once that it yields 1 for lengths, 2 for areas, and 3 for volumes, conformably with Euclidean geometry.

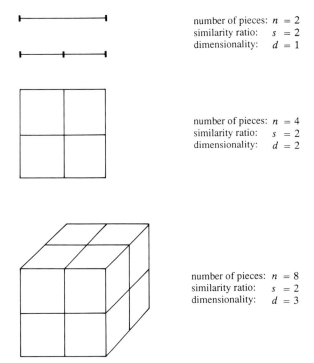

number of pieces: $n = 2$
similarity ratio: $s = 2$
dimensionality: $d = 1$

number of pieces: $n = 4$
similarity ratio: $s = 2$
dimensionality: $d = 2$

number of pieces: $n = 8$
similarity ratio: $s = 2$
dimensionality: $d = 3$

Fig. 6.11 The scaling law $n = s^d$ supplies a concrete realization of the notion of dimensionality.

number of pieces: $n = 2$
similarity ratio: $s = 3$
dimensionality: $d = 0.63$

Fig. 6.12 The Cantor set has fractional dimensionality.

Next, we extend the domain of application of this relation, by using it to determine the dimensionality of a fractal structure, namely of the Cantor set (Fig. 6.12). In going from one to the next higher level, the similarity ratio is $s = 3$, but the number of pieces is $n = 2$. The dimensionality is derived through the relation $d = \ln 2/\ln 3$. Accordingly we find $d \approx 0.63$. This dimensionality is a fraction, whence one speaks of fractal structures, or simply of fractals. We have thus acquired the language to describe quantitatively the geometrical fine structure of figures or of objects, such properties and structure not being limited to the notions only of length, area, and volume. Fractal geometry is the creation of the mathematician Benoît Mandelbrot.

3 Simple realization of a chaotic system

In Section 6.3 we found (at least on a qualitative level) a very simple interpretation of the experiment on the crazed compass needle subject to a fixed plus a rotating field. By contrast, the experimental realization presented considerable practical problems: convenient frequencies of the rotating field are of the order of 1 Hz, and one requires quite a lot of apparatus to produce two sinusoidal B-fields whose phases are in quadrature. At the cost of slightly complicating the interpretation (two rotating and one fixed field) one can make do with far simpler apparatus (Fig. 6.13).

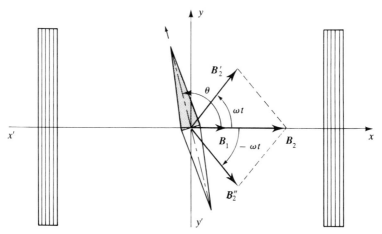

Fig. 6.13 Sketch of an easily set up demonstration experiment to show the chaotic motion of a magnetized needle.

To keep the geometry simple, the axis $x'x$ of the set-up is chosen along the horizontal component B_1 of the Earth's field. On this axis we align two 300-turn coils each having diameter 9 cm; they are wound with copper wire of 0.1 mm diameter. They are connected in series, and fed by a generator supplying alternating current at 1 Hz, of maximum amplitude 70 mA. At the centre of symmetry of the set-up, the coils produce an alternating field B_2 along $x'x$, having frequency 1 Hz and maximum amplitude 2×10^{-4} T. At this centre of symmetry, we place a magnetized needle of classic design, pivoted on a vertical axis; the needle is 7 cm long and its mass is 0.7 g.

The alternating field B_2 along a fixed axis may be considered as the resultant of two rotating fields: B'_2 rotating in the positive direction, with angular velocity ω, and B''_2, rotating in the negative direction, with angular velocity $-\omega$. The magnetic needle is thus subject to three B-fields:

- a fixed field B_1, the horizontal component of the Earth's field, of magnitude 0.2×10^{-4} T;
- two rotating fields, B'_2 and B''_2, of magnitude 2×10^{-4} T.

A touch of the fingers suffices to set the initial conditions; the needle embarks on disordered motion that follows B'_2 some of the time and B''_2 the rest of the time. It is impossible to launch the needle in the same way twice running, because the system's extreme sensitivity to the initial conditions. Thus we have constructed a chaotic system, governed essentially by the action of the fields B'_2 and B''_2, since the effects of B_1 are weak. Other chaotic regimes can be observed by changing the frequency or the intensity of the current.

Finally, the apparatus can be placed on an overhead projector, where it makes a very telling demonstration.

Further reading

See Bergé, Pomeau, Vidal, and Tuckermann (1984), Crutchfield, Farmer, Packard, and Shaw (1986), Ekeland (1988), Lorenz (1963), and Mandelbrot (1982), in the Bibliography.

Niels Bohr. Danish physicist; born and died in Copenhagen (1885–1962). While still a very young man he works with two great British physicsts: Thomson and Rutherford. In 1913 he proposes a planetary model of the atom, which can explain and predict spectral lines; this theory earns him the Nobel prize in 1922. In 1919 he sets up a Laboratory of Theoretical Physics in Copenhagen, which quickly becomes world-famous. In particular, it is there that Heisenberg, Dirac, Pauli, and Ehrenfest elaborate quantum mechanics in the form in which we use it today. As for Bohr, he is both an inspirer and a promoter of the new theory. He contributes greatly to formulating the basic physical significance of quantum mechanics, by propagating the 'Copenhagen interpretation', which amounts to a true conceptual revolution. (Boyer-Viollet)

7

Bohr, or chance unavoidable (quantum mechanics)

Know you her secrets none can utter?
Quiller-Couch

Once more, and for the last time, we cite the classic problem of the gunner (Fig. 7.1).

If a shot is repeated many times, we know that the points of impact will be distributed at random within the scatter zone. We know, further, that by improving the precision with which the initial conditions can be reproduced, we can reduce the area of the zone. We envisage an ideal limit corresponding to initial conditions defined with absolute precision, which would reduce the scatter zone to a point; this is what we call determinism. Now however we

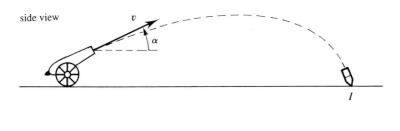

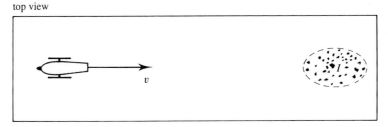

Fig. 7.1 The problem of the gunner.

must face the basic question: what happens if this idealized limit does not exist? In other words, what happens if, before absolute precision is attained, we run into an insuperable bound on precision, due not to our lack of skill, but to physical properties of the projectile in interaction with the gun which fires it? It is obvious that in such a case the scatter would persist, and that the physical theory that furnishes the ultimate description of the motion would have to be probabilistic. In practice this does not worry the gunner, whose projectile is macroscopic. But for the particle physicist the problem is crucial, because the projectiles he uses (photons, electrons, protons, neutrons, etc.) behave in ways that cannot be described within the deterministic framework of classical mechanics. Here a new kind of mechanics is needed, called *quantum mechanics*; developed in stages between 1900 and 1930, it is linked with the greatest names in physics: Max Planck, Albert Einstein, Niels Bohr, Louis de Broglie, Erwin Schrödinger, Werner Heisenberg, and Paul Dirac. The history of this long march is recounted with verve and humour by George Gamow in a book called *Thirty Years that Shook Physics* [10], in view of the fact that quantum mechanics represents a true revolution in the basic concepts of the physicist. Niels Bohr contributed more than anyone else to popularizing the new ideas, especially in a debate with Albert Einstein that started in 1927. Much of this debate will be outlined in Chapter 8. Meanwhile, in presenting quantum mechanics we shall watch our language very carefully, because behind the words there often lurk implicit concepts that, though valid macroscopically, have no currency on the microscopic scale. To give such verbal prudence its due weight, we shall not hesitate on occasion to appeal to the proverbial wisdom of Monsieur de La Palice,† since, through thick and thin, he is the prime advocate of simple-minded truths.

7.1 Single-photon interference

All good books on quantum mechanics start with a long and interesting analysis of the Young's slits interference experiment. The crux of the

---

† The seigneur de La Palice (1471–1525) was a distinguished military commander in the service of Francois I. Killed in the battle of Pavia, he is celebrated in a song:

> Monsieur de La Palice est mort,
> Mort devant Pavie.
> Un quart d'heure avant sa mort
> Il était encore en vie.

(Monsieur de La Palice is dead,/killed at Pavia./Fifteen minutes before his death/he was still very much alive.)

   This was meant to say that he fought to his last breath. But in the course of time the laudatory intent of the lines has got lost, and only the literal meaning has survived. That is how in spite of himself Monsieur de La Palice has become the symbol in France of simple-minded truths.

discussion comes when 'the light intensity is reduced sufficiently for photons to be considered as presenting themselves at the entry slit one by one'. For a long time this point remained contentious, because correlations between two successive photons cannot be ruled out *a priori*. Since 1985 however the situation has changed: P. Grangier, G. Roger, and A. Aspect, at Orsay, performed an interference experiment with truly a single photon. They used a light source devised for an EPR experiment (see Chapter 8), which guarantees that photons arrive at the entry slit singly [11]. The experiment is difficult to do in practice, but it is very simple in principle, and it provides an excellent experimental introduction to the concepts of quantum mechanics.

The light source is a beam of calcium atoms, excited by two focused laser beams having wavelengths $\lambda' = 406$ nm and $\lambda'' = 581$ nm respectively. Two-photon excitation produces a state having the quantum number $J = 0$. When it decays, this state emits two monochromatic photons having the wavelengths $\lambda_1 = 551.3$ nm and $\lambda_2 = 422.7$ nm respectively, in a cascade of two electronic transitions from the initial $J = 0$ state to the final $J = 0$ state, passing through an intermediate $J = 1$ state, as shown in Fig. 7.2.† The mean life of the intermediate state is 4.7 ns. To simplify the terminology, we shall call the $\lambda_1 = 551.3$ nm light green, and the $\lambda_2 = 422.7$ nm light violet.

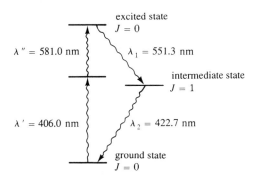

Fig. 7.2 Excitation and decay of the calcium atom.

Next we describe the experiment, exhibiting its three stages which reveal the complications of the apparatus in progressively greater detail (Figs 7.3–7.5).

1. The first stage is a trivial check that the apparatus is working properly; nevertheless it is already very instructive (Fig. 7.3).

† The electronic structures of these states are as follows:
initial $J = 0$ state: $4p^2 1S_0$;
intermediate $J = 1$ state: $4s4p1 P_1$;
final $J = 0$ state: $4s^2 1S_0$.

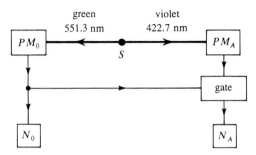

Fig. 7.3 Interference with a single photon (first stage). In this sketch, thick lines show optical paths; thin lines show electrical connections.

On either side of the source S one positions two photomultiplier tubes PM_0 and PM_A. These are very sensitive, and can detect the arrival of a single photon. Detection proceeds through photoelectric absorption, followed by amplification which produces an electric signal proportional to the energy of the incident photon. The associated electronic logic circuits can identify the photons absorbed by each detector: the channel PM_0 responds only to green light, and PM_A only to violet light. The electronic gate is opened (for 9 ns†) when green light is detected by PM_0. If, while the gate is open, violet light is emitted by the same atom towards PM_A,‡ then PM_A detects this photon,§ producing a signal that passes through the gate and is counted in N_A. Finally, there are counters to register the number N_0 of green photons detected by PM_0, and the number N_A of violet photons detected by PM_A while the gate is open, over an observation period of several hours. Obviously N_A is much smaller than N_0, but as the observation period becomes very long the ratio N_A/N_0 tends to a limit that is a characteristic of the apparatus.¶

The purpose of this arrangement is to use a green photon in order to open a 9 ns time window, in which to detect a violet photon emitted by the same atom. The second stage of our discussion will show that there is only a very small probability of detecting through the same window another violet photon emitted by a different atom.

† This window, which is close to twice the mean life, corresponds to an emission probability of 0.85 for the second photon.

‡ The second photon can be detected in any direction, and its angular distribution is anisotropic. The solid angle subtended at the source S by the aperture of PM_A defines the geometrical condition for detection by PM_A.

§ Eventually one must take into account the photoelectric efficiency of the photocathode; this is always less than 1, because the energy of some photons is absorbed without photoemission, contributing only to the heating of the photocathode.

¶ Observation periods are of the order of five hours. Checking the constancy of the ratio N_A/N_0 checks that the apparatus is working properly. In this way one determines an experimental frequency whose limiting (long-period) value corresponds to the probability of detecting a violet photon in PM_A during the 9 ns following the detection of a green photon by PM_0.

So far we have talked the normal language of the physicist. Now it is time to appeal to Monsieur de La Palice. Consulted about this first experiment, he states his views with some self-satisfaction, which does not however preclude linguistic precision, for he uses, one after another, the three formulae 'I observe', 'I conclude', and 'I envisage'.

'I observe that the photomultiplier PM_A detects violet light when the source S is on, and that it ceases to detect anything when the source is off. I conclude that the violet light was emitted by S, and that it travelled from S to PM_A.

I observe that energy is transferred between the light and the photomultiplier PM_A always in the same amount, which I shall call a quantum.

I envisage the quanta as particles, emitted by the source, propagating from S to PM_A, and absorbed by the detector. I shall call these quanta photons.'

Whereupon Monsieur de La Palice retires to his apartments with dignity.

2. The second stage of the experiment introduces the concept of individual photons (Fig. 7.4).

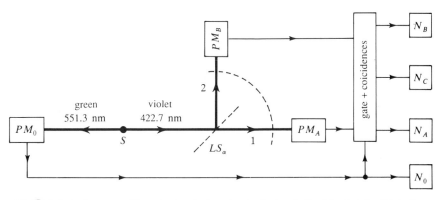

Fig. 7.4 Interference with a single photon (second stage). In this sketch, thick lines show optical paths; thin lines show electrical connections.

Across the path of the violet light one places a half-silvered mirror LS_α, which splits the primary beam into two secondary beams, one transmitted and detected by PM_A, the other reflected and detected by PM_B.† As in the first stage, the gate is opened, for 9 ns, by PM_0. While it is open, one registers detection either by PM_A (counted as N_A); or by PM_B (counted as N_B); or by

---

† If the mirror is chosen and oriented appropriately, then it transmits half and reflects half the light. This is the condition for obtaining high-contrast interference fringes in the third stage of the experiment.

both, which we call a coincidence (counted as N_C). The experiment runs for five hours, and yields the following results.

- The counts N_A and N_B are both of the order of 10^5. By contrast, N_C is much smaller, being equal to 9.
- The sequence of counts from PM_A is random in time,† as in the sequence of counts from PM_B.
- The very low value of N_C shows that counts in PM_A and in PM_B are mutually exclusive.

The experimenters analyse the value of N_C in depth [11]; their reasoning can be outlined as follows.

- Suppose two different atoms each emit a violet photon, one being transmitted to PM_A, and the other reflected to PM_B, with both arriving during the 9 ns opening of the gate; then the circuitry records a coincidence. In the regime under study, and for a run of five hours, quantum theory predicts that the number of coincidences should be $N_C = 9$. The fact that this number is so small means that, in practice, any given single photon is either transmitted or reflected.‡.

- If light is considered as a wave-train, split into two by LS_x and condensed into quanta on reaching PM_A and PM_B, then one would expect the photon counts to be correlated in time, which would entail $N_C \gg 9$. Classically speaking this would mean that there can be no transmitted wave without a reflected wave.

- Experiment yields $N_C = 9$; this quantal result differs from the classical value by 13 standard deviations; hence the discrepancy is very firmly established, and allows us to assert that we are indeed dealing with a source of individual photons.

Monsieur de La Palice leaves such logic-chopping to professionals. Once he notes that N_C is very small, he is quite prepared to treat it as if it were zero. Flattered to be consulted again, he pays close attention, and delivers himself thus:

'I observe that light travels from the source to PM_A or to PM_B, because detection ceases when the source is switched off.
I observe that the counts N_A and N_B correspond to a game of heads or tails,

---

† One check this say by recording the value of N_A every five seconds over the five hours. Then the distribution of the differences ΔN_A, that is of the counts registered in each five-second interval, should be Poissonian (Section 2.5).
‡ Total mutual exclusion between photons would correspond to $N_C = 0$. It could be realized only in the limit of a zero-intensity source. But, in fact, the experimental value $N_C = 9$ automatically implies anticorrelation, without any need for appeal to quantum mechanics.

in that the two possibilities are mutually exclusive, and that the counts are random.

I observe that the optical paths 1 and 2 are distinguishable, because the experiment allows me to ascertain, for each quantum, whether it has travelled path 1 (detection by PM_A) or path 2 (detection by PM_B).

I envisage that, on arrival at the half-silvered mirror, each photon from the source is directed at random either along path 1 or along path 2; and with your permission I will make so bold as to assert that it is in the nature of photons to play heads or tails.'

As well-pleased with himself as ever, Monsieur de La Palice retires again.

3. The third stage consists of an interference experiment (Fig. 7.5). A Mach–Zehnder interferometer is used, allowing one to obtain two interference profiles. The beam of violet light from the source S is split into two by the mirror LS_α. After reflection from two different mirrors, these secondary beams meet on a second half-silvered mirror LS_β. Here, each secondary beam is further split into two; thus one establishes two interference regions, region $(1', 2')$ where one places PM_A, and region $(1'', 2'')$ where one places PM_B.

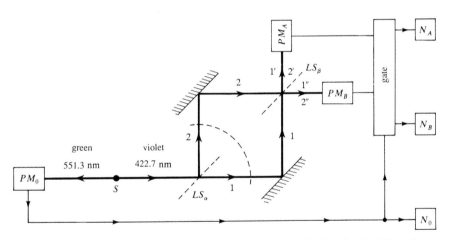

Fig. 7.5 Interference with a single photon (third stage). In this sketch, thick lines show optical paths; thin lines show electrical connections.

A very high precision piezoelectric system allows one of the mirrors to be displaced so as to vary the path difference between the two arms of the interferometer. In this way one can shift the pattern of interference fringes by regular steps, without moving the detectors PM_A and PM_B; the standard step corresponds to a change of $\frac{1}{50}\lambda$ in the difference between the two optical paths.

A sweep, taking 15 s for each standard step, yields two interference plots corresponding, respectively, to the paths (1', 2') and (1'', 2''); the fringes have good contrast, and their visibility† was measured as 98% (Fig. 7.6). Notice that the two interference patterns are staggered by half a fringe-width.‡ This makes it possible to adjust the increments in path-difference so as to realize the most typical case, with an intensity maximum for PM_A and a minimum for PM_B. If we recall that we are reasoning in terms of photons, and that the photons are being processed individually, then we must admit that the interference does not stem from any interaction between successive photons, but that each photon interferes with itself; the adjustment we have adopted produces constructive interference along the path (1', 2'), and destructive interference along (1'', 2''). What would Monsieur de La Palice have to say?

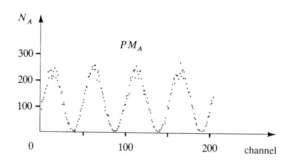

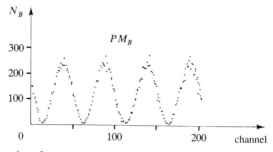

Fig. 7.6 The two interference plots obtained with the Mach–Zehnder interferometer. Note that maximum counting rates in PM_A correspond to minima in PM_B, indicating a relative displacement of $\frac{1}{2}\lambda$ between the two interference patterns.

† Fringe visibility is defined by the ratio

$$(N_{A\ max} - N_{A\ min})/(N_{A\ max} + N_{A\ min}).$$

‡ Their relative displacement stems from the different sequences of reflection-induced phase shifts along the light paths (1', 2') and (1'', 2'').

When Monsieur de La Palice reappears on the scene, one suspects a touch of exasperation under his unvarying courtesy. His contribution is brief:

'I observe that the optical paths differ in length between LS_α and LS_β, and are then coincident over (1′, 2′) and over (1″, 2″).

In PM_A I observe a process that seems perfectly natural to me, namely

$$\text{light} + \text{light} \rightarrow \text{light}.$$

In PM_B I observe a process that I find astounding, namely

$$\text{light} + \text{light} \rightarrow \text{darkness}.$$

Such superposition phenomena with light I shall call interference, constructive in PM_A and destructive in PM_B.

In the situation considered before, I envisaged light as consisting of particles called photons, which travelled either along path 1 or along path 2. In the present situation I want to know for each individual photon which path it has travelled; to this end I should like to ask you close off path 2, since this will ensure that all the photons travel by path 1.'

Clearly, Monsieur de La Palice is perturbed. Instead of withdrawing as usual, he stays put, pretending indifference; but one can see that he awaits the outcome of this new experiment with some anxiety.

On closing either path, whether 2 or 1, one observes that all interference phenomena disappear. For instance, instead of a very high count N_A and a very low count N_B, we now obtain in every case essentially equal counts from PM_A and PM_B.

Visibly displeased and put on his guard by this result, Monsieur de La Palice now pursues his analysis of the experiment thus:

'I observe that in order to produce interference phenomena it is necessary to have two optical paths of different lengths, both open.

Whenever a photon is detected, I note my inability to ascertain whether the light has travelled by path 1 or by path 2, because I have no means for distinguishing between the two cases.†

If I were to suppose that photons travel only along 1, then this would imply that path 2 is irrelevant, which is contrary to what I have observed. Similarly, if I were to suppose that photons travel only along 2, then this would imply that path 1 is irrelevant, which is also contrary to my observations.

If I envisage the source S as emitting particles, then I am forced to conclude that each individual photon travels simultaneously along both paths 1 and 2;

---

† This remark excludes the notion that photons follow trajectories in the sense of classical mechanics.

but this assumption contradicts the results of the previous experiment,† which compelled me to envisage that every photon chooses, at random, either path 1 or path 2.

I conclude that the notion of particles is unsuited to explaining interference phenomena.

I shall suppose instead that the source emits a wave; this wave splits into two at LS_α, and the two secondary waves travel one along path 1 and the other along path 2. They produce interference by mutual superposition on LS_B, constructively in (1′, 2′) and destructively in (1″, 2″). At the far end of (1′, 2′) or of (1″, 2″) I envisage each wave as condensing into particles, which are then detected by the photomultipliers.‡

It seems to me that I am beginning to understand the situation. I envisage light as having two complementary forms: depending on the kind of experiment that is being done, it can manifest itself either as a wave, or as a particle, but never as both simultaneously and in the same place. Thus, in the experiment where the path followed by the light can be ascertained,§ light behaves as a particle between the source S and detection by PM_A or PM_B. By contrast, in the experiment where the path followed by the light cannot be ascertained,¶ light behaves at first like a wave, producing interference phenomena; but it behaves like a particle when, afterwards, it is detected through the photoelectric effect. I conclude that light behaves rather strangely, but nevertheless I have the impression that its behaviour can be fully described once one has come to terms with the idea of wave–particle duality.'

Monsieur de La Palice withdraws slowly, provisionally, and as if rather reluctantly. Is it just that in spite of himself he is impressed by all the novelties he has enunciated, or is he still afraid, deep down, that something more remains to be said?

In fact something does remain to be said, since the problem of causality remains open. Let us look carefully at the experimental layouts in stage 2 (Fig. 7.4) and in stage 3 (Fig. 7.5): we see that they have LS_α in common, and that they differ only beyond some boundary like the broken quadrant drawn downstream from LS_α. We have stated that light behaves like a particle or like a wave depending on whether or not one can ascertain the path it takes through the apparatus; but in the two experiments under consideration, the choice between these alternatives must be decided on LS_α, *before* the light has crossed the crucial boundary, that is at a stage where nothing can as yet distinguish between the two kinds of apparatus, since they differ only beyond the point of decision. It is as if the light 'chose' whether to behave like a wave or

† The second stage, Fig. 7.4.
‡ Essentially by PM_A. Since the contrast is 98%, only very few photons are detected by PM_B.
§ Second stage, Fig. 7.4.
¶ Third stage, Fig. 7.5.

like a particle before 'knowing' whether the apparatus it will pass through will elicit interference phenomena or the photoelectric effect. Hence the question of causality is indeed reopened with a vengeance.

Monsieur de La Palice re-enters abruptly and without waiting to be called; he is disconcerted, and expresses himself rather wearily:

'Originally I supposed that light would behave like a wave or like a particle, depending on the kind of experiment to which it was to be subjected.

I observe that the choice must be made on the half-silvered mirror LS_α, before the light reaches that part of the apparatus where the choice is actually implemented; this would imply that the effect precedes the cause.

I know that both waves and particles obey the principle of causality, that is that the cause precedes the effect.

I conclude that light is neither wave nor particle; it behaves neither like waves on the sea, nor like projectiles fired from a gun, nor like any other kind of object that I am familiar with.

I must ask you to forget everything I have said about this experiment, which seems to me to be thoroughly mysterious.'

Whereupon Monsieur de La Palice withdraws rather sheepishly, giving all those present the impression that he would not be back.

In fact, though his absence has been long, it turns out to have run its course. Monsieur de La Palice reappears quite out of the blue, with a contented smile, and his final statement is not without a touch of malice.

'I observe in all cases that the photomultipliers register quanta when I switch on the light source.

I conclude that "something" has travelled from the source to the detector. This "something" is a quantum object, and I shall continue to call it a photon, even though I know that it is neither wave nor particle.

I observe that the photon gives rise to interference when one cannot ascertain which path it follows; and that interference disappears when it is possible to ascertain the path.

For each detector, I observe that the quanta it detects are randomly distributed in time.

If I repeat the experiment several times under identical conditions, then I observe that the photon counts registered by each photomultiplier are reproducible in a statistical sense.† These counts enable me to determine experimentally, for any kind of apparatus, the probability that a given detector will detect a quantum, and it is precisely such probabilities that constitute the results of experiments.

---

† For example, suppose that in the first and in the second experiments PM_A registers N'_A and N''_A respectively. Then one can predict (Section 2.6) that N''_A has a probability 0.68 of being in the interval $N'_A \pm (N'_A)^{1/2}$.

I assert that the function of a physical theory is to predict the results of experiments.

What I expect from theoretical physicists is a theory that will enable me to predict, through calculation, the probability that a given detector will detect a photon. This theory will have to take into account the random behaviour of the photon, and the absence or presence of interference phenomena depending on whether the paths followed by the light can or cannot be ascertained.

To work, gentlemen! You have my best wishes for your challenging task, which I hope will give you much pleasure!'

And with these weighty and well chosen words, Monsieur de La Palice withdraws, this time for good.

Alerted by the call from Monsieur de La Palice, physicists have indeed worked hard, and the much-desired theory has indeed come to light, constituting quantum mechanics as we now know it. It applies perfectly not only to photons but equally to electrons, protons, neutrons, etc.; in fact it applies to all the particles of microscopic physics. For the last sixty years it has worked to the general satisfaction of physicists. Meanwhile however it has produced two very interesting problems of a philosophical nature.

1. Chance as encountered in quantum mechanics lies in the very nature of the coupling between quantum object and experimental apparatus. No longer is it chance as a matter of ignorance or of incompetence: it is *chance quintessential and unavoidable*.

2. Quantum objects behave quite differently from the familiar objects of our everyday experience: whenever, for pedagogic reasons, one essays an analogy with macroscopic models like waves or corpuscles, one always fails sooner or later, because the analogy is never more than partial. Accordingly, the first duty of the physicist is to force his little grey cells, that is his concepts and his language, into unreserved compliance with quantum mechanics; eventually this will lead him to view the actual behaviour of microsystems as perfectly normal. As to the teacher of physics, her duties are if anything more onerous still, because she must convince the younger generations that quantum mechanics is not a branch of mathematics, but an expression of our best present understanding of physics on the smallest scale; and that, like all physical theories, it is predictive.

7.2 Basic formalism

We shall introduce the elements of quantum mechanics in the form of some simple axiomatics. Physicists have devised a new mathematical tool, the transition amplitude from initial to final state, and it is this amplitude that enables one to calculate the requisite probability.

• For the experiment where the photon travels from the source S to the detector PM_A (Fig. 7.7a), we write the transition amplitude from S to PM_A as

$$\langle \text{photon arriving at } PM_A \,|\, \text{photon leaving } S \rangle.$$

This is a complex number, and it should be read from right to left. We can also write it more symbolically as $\langle f | i \rangle$, which means simply 'transition amplitude from initial to final state'. This symbol, introduced by Dirac, is called a 'bracket'. It can be split into two components: the ket vector $|i\rangle$, representing the photon in the initial state, and the bra vector $\langle f|$, representing the photon in the final state.†

The transition probability from the initial state $|i\rangle$ to the final state $\langle f|$ is the squared modulus of the transition amplitude; hence it reads $|\langle f | i \rangle|^2$.

• If the photon emitted by the source can take either of two paths, and if it is in principle possible to ascertain which path it actually does take (Fig. 7.7b), then there are two transition amplitudes:

$$\langle \text{photon arriving at } PM_A \,|\, \text{photon leaving } S \rangle,$$

$$\langle \text{photon arriving at } PM_B \,|\, \text{photon leaving } S \rangle.$$

They can be symbolized more simply as

$$\langle f_1 | i \rangle, \qquad \langle f_2 | i \rangle.$$

In this case there are two distinct probabilities:

$$|\langle f_1 | i \rangle|^2, \qquad |\langle f_2 | i \rangle|^2.$$

The total probability is their sum:

$$|\langle f_1 | i \rangle|^2 + |\langle f_2 | i \rangle|^2.$$

More generally, we would write

$$|\langle f | i \rangle|^2 = \sum_k |\langle f_k | i \rangle|^2.$$

• If the photon emitted by the source S can take either of two paths, but it is

---

† One speaks of 'vectors' in order to indicate, in a mathematical sense, that bras and kets are elements of Hilbert spaces. These are vector spaces whose dimensionality is denumerably infinite. Discussions of Hilbert spaces will be found in all texts on quantum mechanics [5, 18].

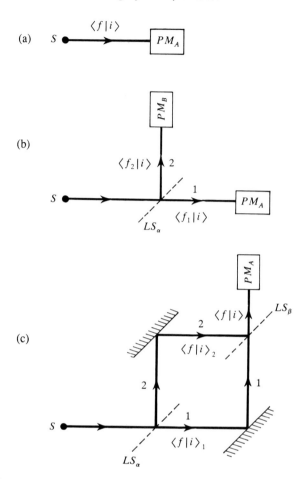

Fig. 7.7 Three arrangements sufficient to determine the transition amplitude: (a) a single optical path; (b) two paths, allowing us to ascertain which path has actually been taken; (c) two paths, not allowing us to ascertain which path has actually been taken.

impossible to ascertain which path it does take (Fig. 7.7c), then there are again two transition amplitudes:

$$\langle \text{photon arriving at } PM_A | \text{photon leaving } S \rangle_{\text{along path 1}},$$

$$\langle \text{photon arriving at } PM_A | \text{photon leaving } S \rangle_{\text{along path 2}}.$$

Symbolically, we write

$$\langle f|i \rangle_1, \qquad \langle f|i \rangle_2.$$

To allow for interference, we assert that in this case it is the amplitudes that must be added; the total amplitude reads

$$\langle f|i \rangle = \langle f|i \rangle_1 + \langle f|i \rangle_2.$$

The total probability is

$$|\langle f|i \rangle|^2 = |\langle f|i \rangle_1 + \langle f|i \rangle_2|^2.$$

More generally, we have

total amplitude: $\quad \langle f|i \rangle = \sum_k \langle f|i \rangle_k;$

total probability: $\quad |\langle f|i \rangle|^2 = \left| \sum_k \langle f|i \rangle_k \right|^2.$

• If one wants to analyse the propagation of the light more closely, one can take into account its passage through the half-silvered mirror LS_α, considering this as an intermediate state (Fig. 7.7b). The total amplitude for path 1 is

$$\langle \text{photon arriving at } PM_A | \text{photon leaving } S \rangle;$$

however, it results from two successive intermediate amplitudes:

$$\langle \text{photon arriving at } LS_\alpha | \text{photon leaving } S \rangle,$$

$$\langle \text{photon arriving at } PM_A | \text{photon leaving } LS_\alpha \rangle.$$

Here we consider the total amplitude as the product of the successive intermediate amplitudes; symbolically, labelling the intermediate state as v, we have

$$\langle f|i \rangle = \langle f|v \rangle \langle v|i \rangle.$$

• Consider finally a system of two mutually independent photons. If photon 1 undergoes a transition from a state i_1 to a state f_1, and photon 2 from i_2 to f_2, then

$$\langle f_1 f_2 | i_1 i_2 \rangle = \langle f_1 | i_1 \rangle \langle f_2 | i_2 \rangle.$$

The four rules just given suffice to calculate the detection probability in any possible experimental situation. They assume their present form as the result of a long historical evolution; but they are best justified *a posteriori*, because in sixty years they have never been faulted. Accordingly, we may consider them as the basic principles governing the observable behaviour of all microscopic objects, that is of objects whose actions on each other are of order h (Planck's constant). From these principles one can derive all the requisite formalism, that is all of quantum mechanics, which we shall introduce through the simple example of the polarization of light.

7.3 The polarization of light

The experiment is very easy to do (Fig. 7.8): one need merely look at a lamp through two sheets of Polaroid, and observe the variation of the transmitted light intensity as one varies the angle between analyser and polarizer. In particular one notes that no light at all is transmitted when the angle between the two sheets of Polaroid is 90°; this is a most astonishing phenomenon, since it amounts to

transparent sheet + transparent sheet → opaque sheet.

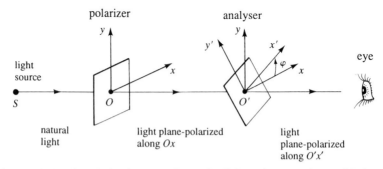

Fig. 7.8 A simple polarization experiment for light, using two sheets of Polaroid.

Quantitative study leads one to identify, in each sheet, two orthogonal axes Ox and Oy. The first axis corresponds to transmission of the incident light with no attentuation at all, and the second to the opposite case, that is to total absorption.† Accordingly we define axes Ox and Oy for the polarizer, and axes Ox' and Oy' for the analyser. The angle $x'Ox$ is called ϕ (Fig. 7.9). Our first object is to determine how transmission varies as a function of ϕ.

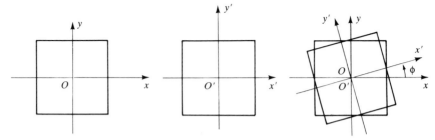

Fig. 7.9 Choice of axes and angles in the experiment on linear polarization. The axes Ox and $O'x'$ correspond, respectively, to the direction of oscillation of the electric field of the light emerging from the analyser and the polarizer.

† At normal incidence, the axis Ox is along the direction of oscillation of the electric field of the emerging light wave, which is linearly polarized.

If the light source is of normal intensity, that is a lamp of the kind used in everyday life, then a photoelectric cell makes a convenient detector; it delivers a continuous current proportional to the intensity of the incident light (Fig. 7.10a). One measures, thus, an intensity I_0 emerging from the polarizer P, and an intensity I' emerging from the analyser A, verifying Malus's law

$$I' = I_0 \cos^2 \phi.$$

Note that this is a law connecting two continuously variable quantities, namely the intensities I_0 and I'; we say that here one is speaking the language of classical physics.

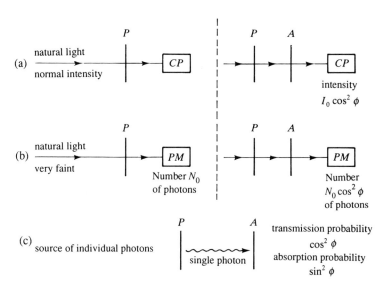

Fig. 7.10 The polarization of light: (a) classical analysis using a photoelectric cell; (b) semi-classical analysis using a photomultiplier; (c) quantum analysis.

If the light source is so weak that it emits individual photons in the sense defined at the start of this chapter, then the appropriate detector is a photomultiplier capable of registering photons one by one (Fig. 7.10b). In this case we count the number N_0 of photons emerging from P over some convenient time interval, and then the number N' emerging from A over an equal interval; Malus's law applies in the form

$$N' = N_0 \cos^2 \phi.$$

Accordingly, we may assert that the number of photons is proportional to the intensity of the light.†

Note that in this case Malus's law connects two discontinuously varying quantities, namely the photon numbers N_0 and N'; we say that the language employed here is semi-classical, since it combines a quantum condition (N proportional to I) with a continuous rule from classical physics ($I' = I_0 \cos^2 \phi$).

Quantum mechanics speaks the language of probabilities. Hence we consider the case of a single photon emerging from P, and therefore polarized along Ox (Fig. 7.10c). We predict two possibilities on its arrival at the analyser A:

- it can emerge from A polarized along Ox, with a transmission probability‡ $\cos^2 \phi$;
- it can fail to emerge from A, with an absorption probability $\sin^2 \phi$.

Note that there are only two possible outcomes, and that they are mutually exclusive; their probabilities must sum to 1, which is indeed what we have in view of

$$\cos^2 \phi + \sin^2 \phi = 1.$$

The quantum formalism adapts to this analysis quite naturally.

The polarizer P defines two states open to the photon, represented by the ket vectors $|x\rangle$ (corresponding to transmission) and $|y\rangle$ (corresponding to absorption); we say that the photon is in the $|x\rangle$, $|y\rangle$ representation, or equivalently that $|x\rangle$, $|y\rangle$ constitute an orthonormal basis. The photon is represented by a state vector $|\Psi\rangle$; if we are interested in the transmitted photon, which is polarized along Ox, then we write the state vector as

$$|\Psi\rangle = 1|x\rangle + 0|y\rangle,$$

or more simply as

$$|\Psi\rangle = |x\rangle.$$

This means that we consider $|x\rangle$ and $|y\rangle$ as unit vectors; we shall see that bras

---

† This is a statistical result, valid to high precision if the numbers N_0 and N' are large; if they are, then the fluctuations $\pm N_0^{1/2}$ and $\pm N'^{1/2}$ are small relative to the numbers themselves. If for example $N_0 = 10^4$, then $N_0^{1/2} = 100$ and $N_0^{1/2}/N_0 = 10^{-2}$; thus the relative deviation is 10^{-2}, at a confidence level of 0.68.

‡ Meaning the probability of detection, beyond A, by a detector of efficiency 1; if the efficiency of the detector is $K < 1$, then the detection probability is $K \cos^2 \phi$.

and kets do indeed have properties very similar to those of vectors in ordinary space.†

The analyser A also defines two states open to the photon, represented by the kets $|x'\rangle$ (corresponding to transmission, with polarization along Ox') and $|y'\rangle$ (corresponding to absorption). Between polarizer and analyser the photon is not subject to any interactions; hence it is represented always by the same state vector $|\Psi\rangle$, which can however be expressed in terms of its components either in the $|x\rangle$, $|y\rangle$ representation, or in the $|x'\rangle$, $|y'\rangle$ representation (Fig. 7.11).

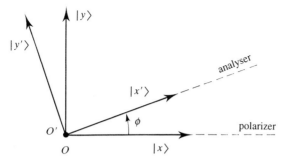

Fig. 7.11 Malus's experiment according to the quantum formalism.

- The expression in the $|x\rangle$, $|y\rangle$ representation is already known:

$$|\Psi\rangle = |x\rangle.$$

- The expression required in the $|x'\rangle$, $|y'\rangle$ representation is found by using the classic formulae‡ for changing the reference frame in vector algebra. Hence we have

$$|\Psi\rangle = \cos\phi\,|x'\rangle - \sin\phi\,|y'\rangle.$$

What we require is the transition amplitude $\langle x'|\Psi\rangle$ for detection in the final state $\langle x'|$. We get it simply by taking the scalar product of both sides of the last equation with the bra vector $\langle x'|$:

$$\langle x'|\Psi\rangle = \cos\phi\langle x'|x'\rangle - \sin\phi\langle x'|y'\rangle.$$

† The only important difference lies in the symmetry properties of the scalar products. In ordinary space, the scalar product of two vectors $\boldsymbol{a}$ and $\boldsymbol{b}$ satisfies the relation $\boldsymbol{a}\cdot\boldsymbol{b} = \boldsymbol{b}\cdot\boldsymbol{a}$; in Hilbert space, the scalar product of the ket vector $|x\rangle$ and the bra vector $\langle x'|$ is a complex number satisfying the relation $\langle x'|x\rangle = \langle x|x'\rangle^*$, where * denotes the complex conjugate [5, 18].

‡ These formulae read

$$|x\rangle = \cos\phi\,|x'\rangle - \sin\phi\,|y'\rangle,$$
$$|y\rangle = \sin\phi\,|x'\rangle + \cos\phi\,|y'\rangle.$$

It follows at once that†

$$\langle x' | \Psi \rangle = \cos \phi, \qquad |\langle x' | \Psi \rangle|^2 = \cos^2 \phi.$$

This shows that the probability for a photon to emerge from A (polarized along Ox') is $\cos^2 \phi$, which is indeed Malus's law.

The approach followed in quantum mechanics stresses the role of the experimental apparatus:

- the polarizer P determines the initial conditions on the photon, namely polarization along Ox;

- the analyser erases all memory of any previous polarization, and defines new initial conditions, namely polarization along Ox'.

These elementary polarization phenomena serve as a very clear manifestation of the most fundamental feature of quantum mechanics (Fig. 7.12).

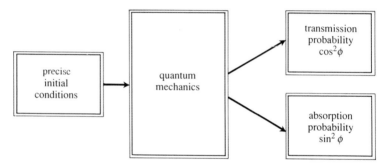

Fig. 7.12 The predictions of quantum mechanics for a simple polarization experiment.

Starting from a precisely defined initial polarization along Ox, quantum mechanics predicts that there are two possibilities for the photon: transmission (probability $\cos^2 \phi$) or absorption (probability $\sin^2 \phi$).

Regarding deterministic theories we have stressed that exact knowledge of the initial conditions allows one to predict the future evolution of the system exactly and uniquely. But in quantum mechanics predictions are probabilistic: given exact initial conditions, we can predict exactly the possible evolutions of the system, attaching a probability to each. Precision is not lost, but the future is no longer unique; this is the basic difference between probabilistic and deterministic theories, between chance and necessity.

† We need merely apply to the bras and kets the rules for constructing scalar products with ordinary vectors: for any unit vector, $\langle x' | x' \rangle = 1$; for the product of two mutually orthogonal unit vectors, $\langle x' | y' \rangle = 0$.

7.4 The machinery of quantum mechanics

It is not our purpose here to deliver a course on quantum mechanics; it would take too long, and there are many books that do it very well. All we intend is to present the basic elements of the quantum formalism, and to explain the physical significance of each element. In this way we hope to demystify the symbols, even if their appearance on the paper remains rather abstract. Readers alarmed by this formalism can safely skip the present section at first reading.

• In classical mechanics, magnitudes describing properties of physical systems are represented by numbers (scalars or components of vectors). In quantum mechanics, any measurable quantity describing a physical property of a particle is represented by a mathematical operator, which we shall denote by A.

• $A|x_k\rangle = a_k|x_k\rangle$ is the *eigenvalue equation* associated with the operator A in the $|x\rangle$ representation. From it one can determine all the eigenstates of the particle in the $|x\rangle$ representation, namely the *eigenvectors* $|x_1\rangle$, $|x_2\rangle$,...., $|x_k\rangle$,.... The same equation yields all the possible results from a physical measurement of the quantity in question: these are given by a_1, a_2,...., a_k,....

• $|\Psi(t)\rangle = \sum_k C_k(t)|x_k\rangle$ corresponds to the principle of spectral decomposition.† At a given time t, and in the $|x\rangle$ representation, the state vector $|\Psi(t)\rangle$ representing the particle is in general a sum of many terms like $C_k(t)|x_k\rangle$. If the measurement is performed at time t, then the probability for finding the particle in the state $|x_k\rangle$ is $|C_k(t)|^2$, in which case the result of the measurement is a_k.

• $i\hbar(\mathrm{d}/\mathrm{d}t)|\Psi(t)\rangle = H|\Psi(t)\rangle$ is the evolution equation, associated with the operator H for the total energy; it is generally called the Schrödinger equation.‡ It is a deterministic equation: the state vector $|\Psi(t_0)\rangle$ fully embodies the initial conditions which, at time t_0, define the initial state exactly,§ and from the equation one can then calculate the state vector $|\Psi(t)\rangle$ which describes¶ the system at time t. By contrast, the act of measurement at

---

† We have simplified by restricting ourselves to denumerable sets of eigenvectors. If the set is continuous, the sum must be replaced by an integral.

‡ The Schrödinger equation governs the propagation of a particle with which one can associate a momentum p and a wavelength λ, related by $p = h/\lambda$. This explains the presence of Planck's constant, either as h, or in its so-called reduced form $\hbar = h/2\pi$.

§ In the sense that the initial conditions determine, exactly, all the components of $|\Psi(t_0)\rangle$ at the time t_0.

¶ In the sense that the evolution equation determines all the components of $|\Psi(t)\rangle$ at the time t.

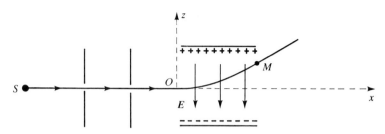

Fig. 7.13 The calculation of the mean trajectory of a beam of charged particles deflected by a uniform electrostatic field.

time t is probabilistic: the principle of spectral decomposition allows us to assert that the probability of finding the result a_k is $|C_k(t)|^2$.

- $\langle \Psi | A | \Psi \rangle$ is the mean value of the physical quantity characterized by the operator A. It is also written $\langle A \rangle$, and it relates to a long sequence of measurements, performed on the same physical system, always under the same conditions specified by the vector $|\Psi\rangle$; afterwards, as the average of the results of all these measurements, one obtains $\langle A \rangle = \langle \Psi | A | \Psi \rangle$.

We proceed to watch this formalism at work in two highly important situations, both very characteristic of quantum mechanics.

7.5 Ehrenfest's theorem

Today it is rather easy to arrange for the deflection of a beam of charged particles, protons or electrons, by a static and uniform electric field; it is done routinely in isotope separation and in cathode-ray tubes. In order to present the theory of this process, we sketch the electron beam in a cathode-ray oscilloscope (Fig. 7.13).

A beam of charged particles is emitted by the source, collimated by two slits T_1 and T_2, and enters the electric field E at the point O, perpendicularly to the field lines. On the scale of our diagram one sees a straight-line incident beam, which is then deflected to describe an arc OM of a parabola while exposed to the field E.

- Generally it is the *classical theory* that is used; it derives from the laws of electromagnetism and of point-particle dynamics, Newtonian or relativistic. The beam is treated as a set of identical particles, each having charge q, mass m, and velocity v. As a first approximation, we take all these particles to have been emitted under the same initial conditions; hence they all describe the same continuous trajectory, which supplies a convenient model of the beam.

To determine the equation of this trajectory, let us write down the components of the basic law of motion (Newton's second law) in the x and in the z directions.

We observe that in the x-direction the particle experiences no force, since the component $\partial V/\partial x$ of the gradient of the electrostatic potential vanishes; therefore the law of motion reads

$$F_x = \frac{\mathrm{d}p_x}{\mathrm{d}t} = -q\,\frac{\partial V}{\partial x} = 0.$$

Since $p_x = mv_x = m\,\mathrm{d}x/\mathrm{d}t$, we have $m\,\mathrm{d}x/\mathrm{d}t^2 = 0$, and, after integration, $x = v_x t$. This is just an example of Newton's first law (the 'law of inertia'), which is a special case of the second law.

The z-component of the electrostatic force reads

$$F_z = qE = -q\,\frac{\partial V}{\partial z} = \text{constant}.$$

The gradient $\partial V/\partial z$ of the electrostatic potential is constant because the field is uniform. Therefore Newton's second law yields

$$F_z = \frac{\mathrm{d}p_z}{\mathrm{d}t} = -q\,\frac{\partial V}{\partial z}.$$

Since $p_z = mv_z = m\,\mathrm{d}z/\mathrm{d}t$, this entails

$$m\,\frac{\mathrm{d}^2 z}{\mathrm{d}t^2} = -q\,\frac{\partial V}{\partial z},$$

or in other words

$$\frac{\mathrm{d}^2 z}{\mathrm{d}t^2} = -\frac{q}{m}\,\frac{\partial V}{\partial z}.$$

Integration leads to

$$z = \tfrac{1}{2}\left(-\frac{q}{m}\,\frac{\partial V}{\partial z}\right)t^2.$$

What we have found is the parametric expression for a parabola:

$$x = v_x t, \qquad z = \tfrac{1}{2}\left(-\frac{q}{m}\,\frac{\partial V}{\partial z}\right)t^2.$$

On our present scale this conclusion from classical theory is well verified experimentally, even though the first section of this chapter showed that the notion of a trajectory makes no sense for elementary particles. Whence comes, then, the success of classical theory in this problem?

• *Quantum theory* answers this question through Ehrenfest's theorem. All the particles in the beam are subject to the same initial conditions. All are acted on in the same way by the electric field, and this action determines the momentum. Given identical initial conditions, every measurement yields a different result; but for the set of all such results we can define average values $\bar{p}_x$ and $\bar{p}_z$, and for these averages Ehrenfest proved the following theorem:

$$\frac{d}{dt}\bar{p}_x = -q\,\frac{\overline{\partial V}}{\partial x}, \qquad \frac{d}{dt}\bar{p}_z = -q\,\frac{\overline{\partial V}}{\partial z}.$$

The parallel with classical theory is immediate: the first equation, where $\partial V/\partial x = 0$, represents Newton's first law; the second equation, where $\partial V/\partial z = $ constant, represents Newton's second law.

Several comments are now in order.

1. In quantum mechanics, the notion of force has limited currency. By contrast, the notion of potential is as useful in quantum as in classical physics.

2. The probabilistic equations of quantum mechanics lead to the deterministic equations of classical mechanics if they are applied to a large number of particles all subject to the same initial conditions. Once again we find circumstances where chance produces determinism.

3. The notion of a trajectory for individual particles makes no sense, but the notion of the mean trajectory is clear and precise: it is an arc of a parabola, calculated from the equations of dynamics in the form they take in Ehrenfest's theorem.

The laws of classical mechanics are the same as those that one has in quantum mechanics for mean values. By assigning to every particle the dynamic characteristics corresponding to such mean values, we achieve a valid classical description of quantum objects. Its validity rests, in the first place, on experiment, in that the electron beam follows the (parabolic) mean trajectory very accurately; but it also rests on Ehrenfest's theorem, which defines the sense in which classical and quantum descriptions correspond.

7.6 Heisenberg's inequalities

These inequalities, established in 1926, are showpieces of quantum mechanics, and they have engendered an abundance of literature both scientific and philosophical. Though we can hardly escape this pattern, we shall watch our language with especial care, because here more than anywhere else one's language is prone to reflect one's preconceptions.

In order to introduce Heisenberg's inequalities step by step, it is best to start from the most characteristic example, that of diffraction of light by a slit. It represents the quantum version of the problem of the gunner, and will serve to bring home the difference between deterministic and probabilistic theories.

The advent of the laser has made diffraction very easy to study, supplying as it does a light beam that is intense, parallel (i.e. highly collimated), and monochromatic. One uses such a beam to illuminate a narrow slit (width of the order of a millimetre), and one observes the distribution of the light on a screen downstream from the slit. Geometric optics would lead one to expect to see a strongly illuminated narrow rectangle, with width equal to the slit-width. What one actually sees is a series of fringes of rather greater width, with one very intense central fringe twice as wide as the others; one says that these fringes constitute a diffraction pattern. If the slit is narrowed, one observes that the fringes spread out; if on the other hand the slit is progressively widened, then the fringe system contracts, reducing eventually to the single illuminated rectangle of geometrical optics. It is this inverse relation between the widths of the slit and of the fringe pattern that characterizes diffraction, and it is the job of physical theory to explain it.

The *classical theory* was proposed by Fresnel in the nineteenth century. The incident light beam is parallel and monochromatic, having wavelength λ, and it falls normally on to a slit of width $2\Delta z$. The diffraction pattern is observed on a screen a distance d downstream from the slit (Fig. 7.14). Strictly speaking we should insert a lens behind the slit, in order to produce at a finite distance the pattern which otherwise would be produced only at infinity. Nevertheless, if d

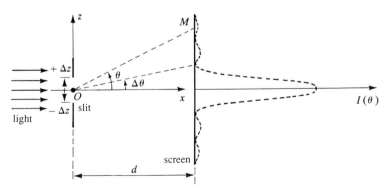

Fig. 7.14 The diffraction of light by a slit: classical analysis.

is large enough compared to the slit-width $2\Delta z$, then the theory of diffraction at infinite distance can be applied to a good approximation, and to good advantage because it is fairly simple. Light is treated as wave-motion, and we calculate the intensity at a point M, defined by the angle θ. Another approximation now becomes possible, provided the angle θ, expressed in radians, is small compared to 1. This validates the approximation $\sin\theta \approx \theta$, which yields a simple expression for the intensity diffracted in the direction θ:

$$I(\theta) = I(0) \left(\frac{\sin\left(\frac{2\pi\Delta z}{\lambda}\theta\right)}{\frac{2\pi\Delta z}{\lambda}\theta} \right)^2 .$$

This expression accounts very well for the observed diffraction pattern, having

- a principal maximum, very luminous since it includes 97% of the total intensity;
- rather faint secondary maxima, whose intensity decreases as one moves away from the principal maximum.

To a good approximation, one can say that all the intensity resides in the central peak, that is between diffracted angles $\Delta\theta$ and $-\Delta\theta$; calculation yields

$$\Delta\theta = \lambda/2\Delta z.$$

This formula expresses the inverse relation between the widths of the slit and of the diffraction pattern, as observed experimentally. Accordingly, Fresnel's theory is well suited to explain diffraction, but it takes into account only the undulatory features of light; in this respect it is incomplete.

The *semi-classical theory* embraces all the results of Fresnel's classical theory, to which it adds one quantum condition: the light beam consists of a set of photons each having momentum p and wavelength λ, related through $p = h/\lambda$ (Fig. 7.15).

The detector is no longer a screen observed with the naked eye. but a photomultiplier tube placed at M, just behind a diaphragm which allows close definition of θ to within $\Delta\theta$. Over a suitably chosen time interval, the detector counts a number $N(\theta)$ of photons. The angle θ can be varied by moving the photomultiplier and its diaphragm up and down; in this way one makes many repeated measurements of $N(\theta)$, each with the same duration. The experimental end-result reads

$$N(\theta) = N(0) \left(\frac{\sin\left(\frac{2\pi\Delta z}{\lambda}\theta\right)}{\frac{2\pi\Delta z}{\lambda}\theta} \right)^2 .$$

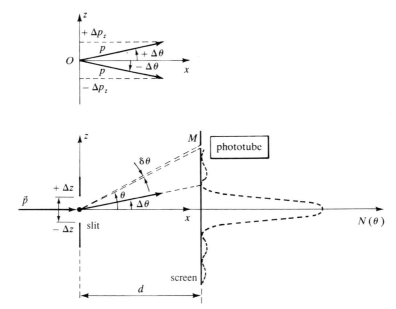

Fig. 7.15 The diffraction of light by a slit: semi-classical analysis.

This empirical expression is in complete accord with the expression given by the Fresnel theory for $I(\theta)$ in terms of $I(0)$; once again we see that the number of photons is proportional to the intensity of the light.

The link with the notion of individual photons is through the probability density

$$\rho(\theta) = \rho(0) \left(\frac{\sin\left(\frac{2\pi\Delta z}{\lambda} \theta\right)}{\frac{2\pi\Delta z}{\lambda} \theta} \right)^2 ,$$

whose significance is the following: for a photon incident on the slit in the x-direction, the probability that it will be detected in the range of angles $\Delta\theta$ around θ is $\rho(\theta)\Delta\theta$. We have here a new and semi-classical version of the gunner's problem: the gun is the laser and the projectile the photon. Even when the gun is pointed exactly in the x-direction, the future of the projectile is still not uniquely determined; on the contrary, there are infinitely many possibilities, specified by the angle θ and the probability density $\rho(\theta)$. The semi-classical theory is probabilistic rather than deterministic.

Our reason for citing the gunner's problem is that the semi-classical theory assumes, implicitly, that the photon can be treated like a point-particle of matter, to which one can at any instant t assign a position and a momentum, and which describes a continuous trajectory from source to detector.

If these assumptions are granted, then one can analyse the future of a photon arriving at the entry slit in the x-direction, with momentum p; for this analysis we deliberately use the word 'uncertainty', even though we shall have to reject it later on. The uncertainty in this case represents our lack of information about the exact position and momentum of the photon; we are assuming that these quantities do exist simultaneously, even if in the present experiment they cannot be measured exactly. Hence our terminology is provisional, linked as it is to the semi-classical theory, and we shall have to refine it in due course. Having said this, we proceed to study the motion in the z-direction.

The uncertainty in the z-coordinate is Δz, because the photon position must be in the interval $\pm \Delta z$ in order to go through the slit and reach the phototube (Fig. 7.15).

The uncertainty in the component p_z of $\boldsymbol{p}$ is Δp_z, which we determine by observing that, essentially, the photon is destined to be diffracted within the central maximum (with a probability greater than 0.97); hence the photon is confined to the angular interval $\pm \Delta \theta$ (Fig. 7.15). It follows that[†]

$$\Delta p_z = p \Delta \theta.$$

Next we appeal to de Broglie's relation

$$p = h/\lambda,$$

where h is Planck's constant, and to Fresnel's result

$$\Delta \theta = \lambda/2\Delta z$$

for the innermost diffraction zero. From them we derive

$$\Delta p_z = \frac{h}{\lambda} \frac{\lambda}{2\Delta z},$$

whence

$$\Delta z \Delta p_z = \tfrac{1}{2} h,$$

a relation obeyed with a probability of 0.97. If we require a relation obeyed with a probability of 1, then we must write the inequality

$$\Delta z \Delta p_z \geq \tfrac{1}{2} h.$$

---

† The calculation assumes that the photon arrives at the centre of the slit. It makes very little difference if it does not, because the condition $d \gg 2\Delta z$ entails that $\Delta \theta$ is practically independent of the entry coordinate of the photon (within $\pm \Delta z$). Incidentally, the calculation yields $\Delta p_z = p \sin \Delta \theta$, but the approximation $\sin \Delta \theta \approx \Delta \theta$ is well warranted.

This rule, written in this form, is the first uncertainty principle, which may be formulated thus:

It is impossible to measure simultaneously the z-components of the position and of the momentum of a particle. There is always an uncertainty Δz in position, and an uncertainty Δp_z in momentum; and the product of these two uncertainties is a constant of order h.

This immediately entails complementarity between the particle and the wave properties of light.

- If Δz decreases, then the localization of the photon improves, and in the limit where Δz reaches 0, one may assert that the photon is a point-object, that is a particle. But at the same time Δp_z becomes infinite; in view of $p = h/\lambda$ this means that the value of λ is wholly unknown, whence all wave properties have disappeared completely.

- If Δp_z decreases, then the definition of the momentum improves; in the limit where Δp_z reaches 0, the wavelength is known exactly, a situation that typifies the wave nature of light. But at the same time Δz becomes infinite, whence all particle properties have disappeared completely.

The semi-classical theory seems well suited to describing diffraction at a single slit, but in the version we have presented it assumes that the photon follows a continuous trajectory. This version fails as soon as one considers interference from two Young's slits. We have seen that each individual photon interferes with itself, but that it is impossible to find out whether it has passed through one slit, or through the other slit, or through both at once; and that this forces us to renounce the notion of a trajectory. Accordingly, the semi-classical theory seems to be merely *ad hoc*, covering just a few phenomena, and we must proceed to the general theory which applies in every possible case.

Quantum theory refuses to describe the photon through quantities inaccessible to measurement. If it is impossible to measure simultaneously and precisely both the position and the momentum, then the reason is that for the photon there do not exist precise and simultaneous values of position and momentum. The ranges $\pm \Delta z$ and $\pm \Delta p_z$ are not uncertainties as to exact values that we happen not to know; rather they represent spreads in properties of the photon as it goes through the slit. Resolved in the z-direction, these amount to a spread of $2\Delta z$ in position and to a spread of $2\Delta p_z$ in momentum. Accordingly, a quantum object has finite extent; moreover it can be modulated by the measuring apparatus in virtue of the relation $\Delta z \Delta p_z \geq \frac{1}{2}h$.

If we want better localization in space, then the spread in momentum increases. This is the quantum analogue of a property well known in optics: if we attempt to isolate a light ray by narrowing the slit, then, far from achieving isolation, we actually increase the width of the diffraction pattern. Nature refuses to comply with the simple model we try to impose on her; rather it is she who forces us to accept her own rules.

We are now in a position to reformulate our sketch of our diffraction experiment. While semi-classically we tried to deal in trajectories, in quantum theory we take into account the widths $\pm\Delta z$ and $\pm\Delta p_z$ which characterize the measurable properties of the photon at the moment of its passage through the slit (Fig. 7.16).

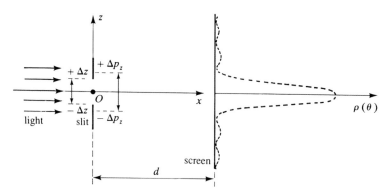

Fig. 7.16 The diffraction of light by a slit: quantum analysis.

The quantum formalism is admittedly more complicated, especially for the photon which is a relativistic particle having speed c. In this book we shall restrict ourselves to non-relativistic quantum mechanics, which excludes photons but which at low energies applies very well to the more familiar particles (electron, proton, neutron, etc.). Since every measurable quantity is specified by an operator, we define, considering z-components,

- the position and momentum operators Q_z and P_z;
- the mean values $\overline{Q_z}$ and $\overline{P_z}$ of the physical quantities represented by these operators;
- deviation operators $\Delta Q_z = Q_z - \overline{Q_z}$
  and $\Delta P_z = P_z - \overline{P_z}$.

One can prove [5] that the mean-square deviations satisfy the relation

$$\overline{\Delta Q_z^2}\ \overline{\Delta P_z^2} \geq \tfrac{1}{4}\hbar^2.$$

From these we define standard deviations Δz and Δp_z through $\Delta z = \left(\overline{\Delta Q_z^2}\right)^{1/2}$ and $\Delta p_z = \left(\overline{\Delta P_z^2}\right)^{1/2}$. Thence one derives the Heisenberg inequality†

---

† There are four such inequalities: three relating to position and momentum, $\Delta x \Delta p_x \geq \tfrac{1}{2}\hbar$, $\Delta y \Delta p_y \geq \tfrac{1}{2}\hbar$, $\Delta z \Delta p_z \geq \tfrac{1}{2}\hbar$; and one relating to time and energy, $\Delta E \Delta t \geq \tfrac{1}{2}\hbar$.

$$\Delta z \Delta p_z \geq \tfrac{1}{2}\hbar,$$

where $\hbar = h/2\pi = 1.054\,680\,8(79) \times 10^{-34}$ J s is (the 'reduced' version of) Planck's constant.

We indicate some experimental situations where this formula applies.

1. Envisage an intense beam of identical particles, all subject to the same initial conditions. We can measure the z-coordinate of each particle very accurately, say by detecting the particle with a photomultiplier placed behind a very narrow slit. The results are distributed at random according to some statistical distribution $f(z)$, with average $\overline{Q_z}$ and standard deviation $\Delta z = (\overline{\Delta Q_z^2})^{1/2}$.

2. Using the same beam, we can measure the z-component p_z of the momentum of each particle very accurately, say by deflecting the particles in a B-field (supposing them to be charged). These results too are distributed at random, according to a different statistical distribution $g(p_z)$, with average $\overline{P_z}$ and standard deviation $\Delta p_z = (\overline{\Delta P_z^2})^{1/2}$.

Between the value of Δz determined by measurements of type 1, and the value of Δp_z determined by measurements of type 2, there holds the relation

$$\Delta z \Delta p_z \geq \tfrac{1}{2}\hbar,$$

which is a law of nature well established both experimentally and theoretically. It can be used to determine Δp_z if one knows Δz, or conversely to determine Δz if one knows Δp_z, thus economizing on measurements.†

3. In simultaneous measurements of the z-components of the position and momentum of a single particle, the results one obtains consist of two pass-bands having widths $2\Delta z$ and $2\Delta p_z$ respectively, related through $\Delta z \Delta p_z \geq \tfrac{1}{2}\hbar$; they represent spreads in the position and momentum of the particle at the instant of measurement. This is the kind of experimental situation that exhibits the deepest physical significance of the Heisenberg inequality, which we have

---

† The functions $f(z)$ and $g(p_z)$ are Fourier transforms of each other; simultaneous knowledge of both leads to Heisenberg's inequality. Actually it is not always a simple matter to exploit the general definition of the deviations Δ_z and Δp_z in terms of the operators ΔQ_z and ΔP_z, but the requisite mathematics does become simple if $f(z)$ and $g(p_z)$ are Gaussians, corresponding to wave-packets (of a special kind). By contrast, for diffraction at a slit Δp_z turns out to be given by a divergent integral if the slit is modelled as a square-step signal, that is as an opening with sharp edges. It is possible to adopt a more realistic model with smoothed and finitely sloping contours, but then the calculations become rather elaborate. One might, of course, fall back onto the conclusion $\Delta z \Delta p_z \geq \tfrac{1}{2}\hbar$ found semi-classically; this method is only approximate, but it does yield a quick order-of-magnitude estimate of the end-result.

On the other hand, one might well be surprised at meeting $\tfrac{1}{2}\hbar$ in some cases and $\tfrac{1}{2}h$ in others. Physically, what is important is to derive a value that is small (of order h) but not zero. Beyond this (as regards factors $1/2$, $1/2\pi$, or $1/4\pi$), everything depends on arbitrary choices made in defining the deviations Δz and Δp_z. The quantum definition in terms of the operators ΔQ_z and ΔP_z is the most widespread, and it is this choice that yields $\tfrac{1}{2}\hbar$.

asserted here in a form designed to replace the traditional formulation of the 'uncertainty principle' in the semi-classical theory. In fact it is high time to forget the word 'uncertainty', substituting for it words like width, dispersion, spread, pass-band, and so on. There is ample choice, and the only essential is to understand that the time for uncertainty has passed!

7.7 Assessment of the model

The credit side in this assessment is considerable, since quantum mechanics utterly dominates microscopic physics, that is most of modern physics as a whole. Rather than cite long lists of successes, we concentrate on presenting a brief synthesis of quantum concepts.

Measurement of a given physical quantity proceeds through interaction between microscopic object and macroscopic measuring apparatus. This interaction is random by nature; the several possible results of the measurement are given by the eigenvalues $a_1, a_2, \ldots, a_k, \ldots$, solutions of the eigenvalue equation

$$A|x_k\rangle = a_k|x_k\rangle,$$

where A is the operator specifying the quantity being measured (Fig. 7.17).

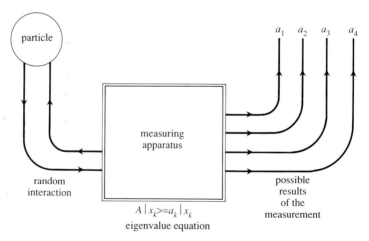

Fig. 7.17 A measurement as described by quantum mechanics.

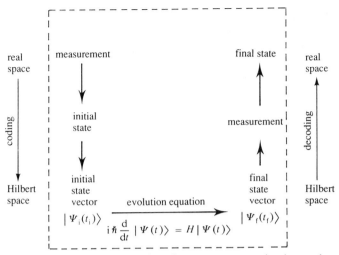

Fig. 7.18 Flow diagram to show how quantum mechanics works.

The time-evolution of initial into final state is fully codified (Fig. 7.18). A measurement at the starting time t_i amounts to defining the initial conditions; it enables one to write down the state vector that describes the system at the time t_i:

$$|\Psi(t_i)\rangle = \sum_k C_k(t_i)|x_k\rangle.$$

The system as it evolves in time is represented by the state vector $|\Psi(t)\rangle$ in Hilbert space. This evolution is deterministic; it can be calculated from the Schrödinger equation

$$i\hbar\frac{d}{dt}|\Psi(t)\rangle = H|\Psi(t)\rangle.$$

At time t_f a second measurement leads to the final state. This operation is probabilistic. The possible results are the eigenvalues $a_1, a_2, \ldots, a_k, \ldots$. The corresponding probabilities are $|C_1(t_f)|^2, |C_2(t_f)|^2, \ldots, |C_k(t_f)|^2, \ldots$. Only one of these possible results is realized, designated conventionally by the index k, giving the result a_k in the final state. In this way one proceeds from a probability $|C_k(t_f)|^2$ before the measurement to a probability 1, that is to certainty, after the measurement. For historical reasons this step is called a collapse (or reduction) of the wave-packet. It is equivalent to the draw in a lottery. Before the draw, a lottery ticket represnts a possible gain, whose probability is low; having been drawn, it represents sure gain.

Finally we note a feature already familiar from our study of chaotic systems, namely a passage from disorder to order achieved by passage from real space to some fictitious space. For chaotic systems this fictitious space is phase space, and the chaos in the real physical system is modelled by a strange attractor which is perfectly orderly. In quantum mechanics one goes from the real, and random, physical system to the state vector in Hilbert space, which behaves deterministically. Thus, quantum mechanics is seen to encode the real world so as to enable us to make predictions about the future, which is the true aim of theoretical physics.

Quantum mechanics, as we have just described it, works splendidly, like a very well-oiled machine. It, and its basic principles, might therefore be expected to command the assent of every physicist; yet it has evoked, and on occasion continues to evoke, reservations both explicit and implicit. For this there are two reasons.

- Quantum mechanics introduces chance unavoidable, meaning that its characteristic randomness is inherent in the microscopic phenomena themselves.

- It attributes to microscopic objects properties so unprecedented that we cannot represent them through any macroscopic analogies.

Both features are revolutionary, and it is natural that they should have provoked debate. On opposite sides of this debate we find two great physicists, Niels Bohr and Albert Einstein, and the next chapter will show how the debate evolved from its beginnings in 1927 to its conclusion in 1983.

Further reading

See Cohen-Tannoudji, Diu, and Laloé (1977), Feynman, Leighton, and Sands (1963–65), Gamow (1985), Grangier, Roger, and Aspect (1986), Hughes (1981), and Robinson (1986) in the Bibliography.

Albert Einstein. Most famous of all physicists; born in Ulm in Germany, died in Princeton in the United States (1879–1955). The vicissitudes of the times lead him to change his nationality more than once: he is successively German, Swiss, German again, and finally American. For the man in the street he is immortal as the creator of the theory of relativity, special (1905) and general (1916). But equally his are the theories of the photoelectric effect (1905), of Brownian motion (1905), of the specific heat of solids (1905), and of the stimulated emission of light (1922). Collaborating with Boris Podolsky and Nathan Rosen in 1935, he envisages the famous EPR paradox, a very searching test of quantum mechanics, addressed chiefly to his adversary and friend Niels Bohr. Over fifty years will elapse before this test is performed and the question settled. (Palais de la Découverte)

8

Inseparable photons (the EPR paradox)

> *The grey-headed parrots whose scientific name is* Agapornis *are more commonly called 'inseparable'. This refers to the behaviour of the couples, who pair for life, and perch together in the closest possible contact.*
>
> (Ornithology)

Though ornithologists have known about inseparable parrots for a long time, to physicists the existence of inseparable photons has been brought home only recently, through a very beautiful series of experiments by Alain Aspect and his group at Orsay. The experiments are exemplary, in virtue both of the difficulties they had to overcome and of the results achieved, which are exceptionally clear-cut. In fact the significance of the experiments extends beyond the strict confines of physics, because they provide the touchstone for settling a philosophical debate that has divided physicists for over sixty years. This division dates back to the appearance of two mutually contradictory interpretations of quantum mechanics at the Como conference in 1927. To sketch the debate, we start with a brief summary of the philosophy of physics.

8.1 The philosophical stakes in the debate

Our summary is best presented diagrammatically (Fig. 8.1).

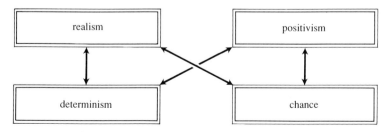

Fig. 8.1 The philosophical elements in a debate between physicists.

- For the physicist who is a *realist*, a physical theory reflects the behaviour of real objects, whose existence is not brought into question.

- For the physicist who is a *positivist*, the purpose of a physical theory is to describe relations between measurable quantities. The theory does not tell one whether anything characterized by these quantities really exists, nor even whether this question makes sense.

- For the physicist who is a *determinist*, exact knowledge of the initial conditions and of the interactions allows the future to be predicted exactly. Determinism is held to be a universal characteristic of natural phenomena, even of those about which we know, as yet, little or nothing. In this framework, any recourse to chance merely reflects our own ignorance.

- For the physicist who is a *probabilist*, chance is inherent in the very nature of microscopic phenomena. To him, determinism is a consequence, on the macroscopic level, of the laws of chance operating on the microscopic level; it is appropriate to measurements of mean values of quantities whose relative fluctuations are very weak.

From these four poles, realism, positivism, determinism, and chance, the physicist chooses two, one on each axis. Though sometimes the choice is made in full awareness of what it entails, most often it is made subconsciously. In our earlier description of quantum mechanics, we adopted without reservations the point of view of the experimentalist working in elementary-particle physics. For a start, he believes firmly in the existence of these particles, since he spends his time in accelerating, deflecting, focusing, and detecting them. Even though he has never seen or touched them, to him their objective existence is not in any doubt. Next he observes that they impinge on the detectors quite erratically, whence he has no doubts, either, that their behaviour is random. Accordingly, the elementary-particle experimentalist has chosen realism and chance, most often without realizing that he has made choices at all.

There are other philosophical options that can be adopted with eyes fully open: realism and determinism are the choices of Albert Einstein; positivism and chance those of Niels Bohr. They are well acquainted and each thinks very highly of the other: which is no bar to their views being incompatible, nor to the two men representing opposite poles of the debate.

8.2 From Como to Brussels (1927–30)

On 26 September 1927, in Como, Niels Bohr delivers a memorable lecture. His stance is that of an enthusiastic champion of the new quantum mechanics. He puts especial weight on the inequalities proved by Heisenberg the year before:

- the position–momentum inequality $\Delta x \Delta p_x \geq \frac{1}{2}\hbar$;
- the time–energy inequality $\Delta t \Delta E \geq \frac{1}{2}\hbar$.

They imply that it is impossible to define exact initial conditions for a microscopic object, which automatically makes it impossible to construct on the microscopic scale a deterministic theory patterned on classical mechanics. Only a probabilistic theory is possible, and that theory is quantum mechanics.

Einstein disagrees with this point of view, and his opposition to Bohr's theses becomes public at the Brussels conference in 1930: he adopts the role of a dissenter who knows precisely how to press home the most difficult questions. Deeply shocked by the retreat from determinism, he tries to show that his famous thought experiments can contravene the Heisenberg inequalities.

At the cost of several sleepless nights devoted to analysing the objections of his adversary, Bohr refutes all of Einstein's criticisms, and emerges from the conference as the undoubted winner.

8.3 From Brussels to the EPR paradox (1930–5)

Having lost the argument at Brussels, Einstein tries to define his objections with ever greater precision. Believing as he does that position and momentum exist *objectively and simultaneously*, he considers quantum mechanics to be incomplete and merely provisional. The points of view of the two antagonists at this stage of the debate can be spelled out as follows.

For Einstein, a physical theory must be a deterministic and a complete representation of the objective reality underlying the phenomena. It features known variables that are observable, and others, unknown as yet, called *hidden variables*. Because of our provisional ignorance of the hidden variables, matter at the microscopic level appears to us to behave arbitrarily, and we describe it by means of a theory that is incomplete and probabilistic, namely by quantum mechanics.

For Bohr, a physical theory makes sense only as a set of relations between observable quantities. Quantum mechanics supplies a correct and complete description of the behaviour of objects at the microscopic level, which means that the theory itself is likewise complete. The observed behaviour is probabilistic, implying that chance is inherent in the nature of the phenomena.

Between chance as a matter of ignorance, as advocated by Einstein, and chance unavoidable, as advocated by Bohr, the debate does not remain merely philosophical. Quite naturally it returns to the plane of physics with the thought experiment proposed in 1935 by Einstein, Podolsky, and Rosen, which in their view proves that quantum mechanics is indeed incomplete. Their thought experiment is published as a paper in the *Physical Review*, but it is so important that it reverberates as far as the *New York Times*. Physicists call the proposal the EPR paradox, after its proponents. It will take fifty years to untangle the question, first in theory and then by experiment. We shall not, of course, follow these fifty years blow by blow; instead, we confine attention to three decisive stages reached respectively in 1952, 1964, and 1983. But we start with an illustration that helps one to see just what the EPR paradox actually is.

8.4 An elementary introduction to the EPR paradox

Consider two playing cards, one red and one black (Fig. 8.2). An experimenter in Lyons puts them into separate envelopes which he then seals. He is thus provided with two envelopes looking exactly alike, and he puts both into a box. He shakes the box so as to 'shuffle the pack', and the system is ready for the experiment.

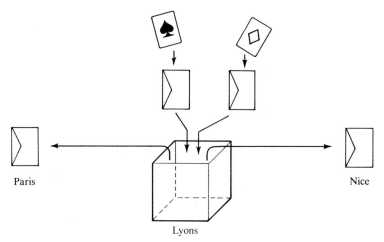

Fig. 8.2 Two playing cards help us understand the stakes in the EPR paradox.

At 8.00 hours two travellers, one from Paris and one from Nice, come to the box (in Lyons), take one envelope each, and then return to Paris and to Nice respectively. At 14.00 hours they are back at their starting points; each opens his envelope, looks at the card, and telephones to Lyons reporting the colour. The experiment is repeated every day for a year, and the observer in Lyons keeps a careful record of the results. At the end of the year the record stands as follows.

1. The reports from Paris are 'red' or 'black', and the sequence of these reports is random. The situation is exactly the same as in a game of heads or tails, and the probability of each outcome is $\frac{1}{2}$.

2. The reports from Nice are 'red' or 'black', and the sequence of these reports is random. Here too the probability of each outcome is $\frac{1}{2}$.

3. When Paris reports 'red', Nice reports 'black'; when Paris reports 'black', Nice reports 'red'. One sees that there is perfect (anti) correlation between the report from Paris and the report from Nice.

Accordingly, the experiment we have described displays two features:

- it is *unpredictable* and thereby random at the level of individual observations in Paris or in Nice;

- it is *predictable*, by virtue of the correlation, at the level where one observes the Paris and the Nice results simultaneously.

Einstein and Bohr might have interpreted the correlation as follows.

According to Einstein, the future of the system is decided at 8.00 hours, when the envelopes are chosen, because he believes that the contents of the two envelopes differ. Suppose, for instance, that Paris has (without knowing it) drawn the red card, and Nice the black. The colours so chosen exist in reality, even though we do not know them. The two cards are moved, separately, by the travellers between 8.00 and 14.00 hours, during which time they do not influence each other in any way. The results on opening the envelopes read 'red' in Paris and 'black' in Nice. Since the choice at 8.00 hours was made blind, the opposite outcome is equally possible, but the results at 14.00 hours are always correlated (either red/black or black/red). This correlation at 14.00 hours is determined by the separation of the colours at 8.00 hours, and we say that the theory proposed by Einstein is *realist, deterministic, and separable,*† by virtue of a hidden variable, namely of colour.

According to Bohr, there is a crucial preliminary factor, inherent in the preparation of the system. On shaking the box containing the two envelopes, one loses information regarding the colours. Afterwards, one knows only that each envelope contains either a red card (probability $\frac{1}{2}$) or a black card

---

† Often such a theory is called realist, deterministic, and local.

(probability $\frac{1}{2}$). We will therefore say that a given envelope is in a 'brown state', which is a superposition of a red state and of a black state having equal probabilities. At 8.00 hours the two envelopes are identical: both are in a 'brown state', and the future of the system is still undecided. There is no solution until the envelopes are opened at 14.00 hours, since it is only the action of opening them that makes the colours observable. The result is probabilistic. There is a probability $\frac{1}{2}$ that in Paris the envelope will be observed to go from the 'brown state' to the red, while the envelope in Nice is observed to go from the 'brown state' to the black; there is the same probability $\frac{1}{2}$ of observing the opposite. But the results of the observations on the two envelopes are always correlated, which means that there is a mutual influence between them, in particular at 14.00 hours; in fact it is better to say that, jointly, they constitute but a single and non-separable system, even though one is in Paris and the other in Nice. Accordingly, the theory proposed by Bohr is *positivist, probabilistic, and non-separable,*† interrelating as it does the colours that are actually observed.

Einstein's view appears to be common sense, while it must be admitted that Bohr's is very startling; however, the point of this macroscopic example is, precisely, to stress how very different the quantum view is from the classical.

Proceeding with impeccable logic but from different premises, both theories predict the same experimental results. Can we decide between them? At the level considered here it seems that we cannot: for even if the envelopes were opened prematurely while still in Lyons, one would merely obtain the same results at a different time, and without affecting the validity of either interpretation. The solution to the problem must be looked for at the atomic level, by studying the true EPR set-up itself.

8.5 The EPR paradox (1935–52)

Albert Einstein, Boris Podolsky, and Nathan Rosen meant to look for an experiment that could measure, indirectly but simultaneously, two mutually exclusive quantities like position and momentum. Such results would contravene the predictions of quantum mechanics, which allows the measurement of only one such quantity at any one time; that is why the thought experiment is called the EPR paradox.

In 1952 David Bohm showed that the paradox could be set up not only with continuously varying quantities like position and momentum, but also with discrete quantities like spin. This was the first step towards any realistically conceivable experiment. Meanwhile, objectives have evolved, and nowadays

† Often such a theory is called positivist, non-deterministic, and non-local.

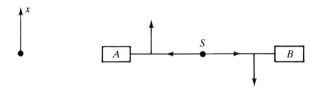

Fig. 8.3 The simplest EPR scenario.

it is more usual to talk of the EPR *scenario*, meaning some sensible experiment capable of discriminating between quantum theory and hidden-variable theories. Such a set-up is sketched in Fig. 8.3.

A particle with spin 0 decays, at S, into two particles of spin $\frac{1}{2}$, which diverge from S in opposite directions. Two Stern–Gerlach type detectors A and B measure the x-components of the spins. Two types of response are possible:

- 'spin up' at A, 'spin down' at B, a result denoted as $(+1, -1)$;
- 'spin-down' at A, 'spin up' at B, denoted as $(-1, +1)$.

This far everyone is agreed, but the interpretation is yet to come.

Einstein reasons that if pairs of particles produced at S elicit different responses $(+1, -1)$ and $(-1, +1)$ from the detector system A, B, then the pairs must have differed already at S, immediately after the decay. It must be possible to represent this difference by a hidden variable λ, which has an objective meaning, and *which governs the future of the system*. After the decay the two particles separate without influencing each other any further, and eventually they trigger the detectors A and B.

Bohr reasons that all the pairs produced at S are identical. Each pair constitutes a non-separable system right up to the time when the photons reach the detectors A and B. At that time we observe the response of the detectors, which is probabilistic, admitting two outcomes, $(+1, -1)$ and $(-1, +1)$.

To sum up, Einstein restricts the operation of chance to the instant of decay (at S), whose details we ignore, but which we believe creates pairs whose hidden variables λ are different. By contrast, Bohr believes that chance operates at the instant of detection, and that it is inherent in the very nature of the detection process: this is chance unavoidable. We are still in the realms of thought, and stay there up to 1964.

In 1964 the landscape changes: John Bell, a theorist at CERN, shows that it is possible to distinguish between the two interpretations experimentally. The test applies to the EPR scenario; it is refined by Clauser, Horne, Shimony, and Holt, whence it is called the BCHSH inequality after its five originators.

8.6 The BCHSH inequality (1964)

To set up an EPR scenario, one first needs a source that emits particle pairs. Various experimental possibilities have been explored:

- atoms emitting two photons in cascade;
- electron–positron annihilation emitting two high-energy photons;
- elastic proton–proton scattering.

It is the first-mentioned solution that has eventually proved most convenient; it has been exploited by Alain Aspect in particular.

Next one needs detectors whose response can assume one of two values, represented conventionally as $+1$ and -1. Such a detector might be

- for spin $\frac{1}{2}$ particles, a Stern–Gerlach apparatus responding 'spin up' or 'spin down';
- for photons, a polarizer responding 'parallel polarization' or 'perpendicular polarization'.

Our sketch of the EPR scenario can now be completed as in Fig. 8.4.

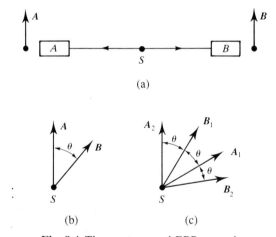

Fig. 8.4 The most general EPR scenario.

Figure 8.4a views the apparatus perpendicularly to its axis, showing the two detectors A and B, with their polarizing directions designated as $\vec{A}$ and $\vec{B}$.

Figure 8.4b views the apparatus along its axis, and shows that the analysing directions of the two detectors are not parallel, but inclined to each other at an angle θ.

Figure 8c is likewise a view along the axis of the apparatus, and shows the actual settings chosen by Aspect: two orientations are allowed for each detector, $\vec{A}_1$ or $\vec{A}_2$ for one, and $\vec{B}_1$ or $\vec{B}_2$ for the other.

We adopt the following conventions:

- $\alpha = \pm 1$ is the response of detector A when oriented along $\vec{A}$:
- $\beta = \pm 1$ is the response of detector B when oriented along $\vec{B}$.

Since each detector has two possible orientations, called 1 and 2, we shall denote their responses as a_1, α_2 and β_1, β_2 respectively.

Now consider the quantity $\langle \gamma \rangle$ defined by

$$\langle \gamma \rangle = \langle \alpha_1 \beta_1 \rangle + \langle \alpha_2 \beta_2 \rangle + \langle \alpha_2 \beta_1 \rangle - \langle \alpha_2 \beta_2 \rangle$$

where the symbol $\langle \; \rangle$ denotes the mean value over very many measured events. We call $\langle \gamma \rangle$ the *correlation function* of the system.

The BCHSH inequality reads

$$-2 \leq \langle \gamma \rangle \leq 2.$$

Its authors have proved that it must be satisfied if mechanics at the microscopic level constitutes a theory that is realist, deterministic, and separable: or in other words if the theory contains a hidden variable (see Appendix 1).

Otherwise, that is according to quantum mechanics (which is positivist, probabilistic, and non-separable), there are cases where the BCHSH inequality is violated. In particular, one can show that for photons in the configuration chosen by Aspect quantum mechanics yields

$$\langle \gamma \rangle = 3 \cos 2\theta - \cos 6\theta.$$

This leads to values well outside the interval $[-2, 2]$, for example to $\langle \gamma \rangle = 2\sqrt{2}$ when $\theta = 22.5°$, and to $\langle \gamma \rangle = -2\sqrt{2}$ when $\theta = 67.5°$ (see Appendix 2).

The BCHSH test turns the EPR scenario into an arena for rational confrontation between the two interpretations; it remains only to progress from thought experiments to experiments conducted in the laboratory.

8.7 From theory to experiment (1964–72)

Though Bell's inequality dates back to 1964, the experiments are so difficult that results started to become available only in 1972. Many different kinds of correlations have been studied; a list (not exhaustive) reads

Berkeley	1972	photons from calcium;
Harvard	1973	photons from mercury-198;
Catagna	1974	γ-rays from e^+e^- annihilation;
Columbia	1975	γ-rays from e^+e^- annihilation;
Berkeley	1976	photons from mercury-202;
Saclay	1976	elastic proton–proton scatter.

All these experiments except Catagna indicate violations of Bell's inequality, but they are so difficult and the fluctuations are so large that the violations barely exceed one standard deviation. Because the stakes are so high, both in physics and in philosophy, at this stage a truly decisive experiment is still lacking; it will be performed by Aspect and his team at Orsay, between 1976 and 1983.

8.8 The beginnings of the experiment at Orsay (1976)

Alain Aspect's experiment studies the correlation between the polarizations of the members of photon pairs emitted by calcium. The light-source is a beam of calcium atoms† excited by two focused laser beams having wavelengths λ' and λ''. Two-photon absorption produces an excited state having $J=0$, which then decays to the $J=0$ ground state in two stages, via a $J=1$ intermediate state. Two photons are emitted in this process, having wavelengths $\lambda_1 = 551.3$ nm and $\lambda_2 = 422.7$ nm respectively. The mean life of the $J=1$ intermediate state is 4.7 ns (Fig. 8.5).

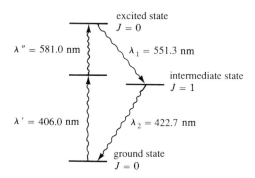

Fig. 8.5 Excitation and decay of a calcium atom.

† Aspect and his group built the source for studying photon correlations in an EPR scenario. Later it was used again to study single-photon interference with a Mach–Zender interferometer (Section 7.1).

The polarizer, which works like a Wollaston prism (Fig. 8.6), is made of quartz or of calcite. It splits an incident beam of natural (unpolarized) light into two beams of equal intensity, polarized at 90° to each other. If only a single unpolarized photon is incident, it emerges either in the state $|x\rangle$, with probability $\frac{1}{2}$, or in the state $|y\rangle$, also with probability $\frac{1}{2}$. Thus the response of the system is two-valued.

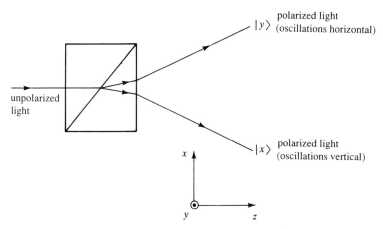

Fig. 8.6 The two-valued response of a Wollaston prism.

The photon is detected by photomultiplier tubes (*PM*) downstream from the prism. Every electric pulse from these detectors corresponds to the passage of a photon, allowing the photons to be counted. The layout is sketched in Fig. 8.7. It uses a coincidence circuit which registers an event whenever two photons are detected in cascade.

In this way four separate counts are recorded simultaneously, over some given period of time:†

- N_{++}, the numbers coincidences corresponding to $\alpha = 1$ and $\beta = 1$, i.e. to $\alpha\beta = 1$;
- N_{+-}, the number of coincidences corresponding to $\alpha = 1$ and $\beta = -1$, i.e. to $\alpha\beta = -1$;
- N_{-+}, the number of coincidences corresponding to $\alpha = -1$ and $\beta = 1$, i.e. to $\alpha\beta = -1$;

† In the EPR scenario envisaged by Bohm, where $\theta = 0$, the only possible responses are $(+1, -1)$ or $(-1, +1)$. In the situation realized by Aspect, the angle θ is non-zero, and four different responses are possible.

- N_{--}, the number of coincidences corresponding to $\alpha = -1$ and $\beta = -1$, i.e. to $\alpha\beta = 1$.

The resolving time of the coincidence circuit is 10 ns, meaning that it reckons two photons as coincident if they are separated in time by no more than this. The mean life of the intermediate state of the calcium atom is 4.7 ns. Hence, after a lapse of 10 ns, that is of more than twice the mean life, almost all the atoms have decayed (specifically: 88%). In other words the efficiency of the coincidence counter is rather high.

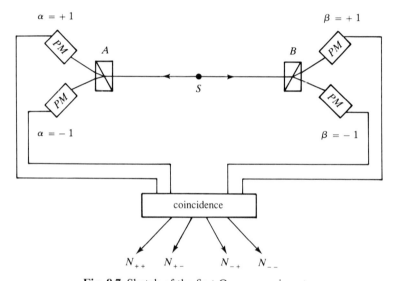

Fig. 8.7 Sketch of the first Orsay experiment.

The experiment consists in counting, over some given time interval, the four kinds of coincidence: N_{++}, N_{+-}, N_{-+}, and N_{--}. The total number of events is

$$N = N_{++} + N_{+-} + N_{-+} + N_{--}.$$

Accordingly, the different kinds of coincidence have probabilities

$$
\begin{aligned}
P_{++} &= N_{++}/N, &&\text{corresponding to } \alpha\beta = 1,\\
P_{+-} &= N_{+-}/N, &&\text{corresponding to } \alpha\beta = -1,\\
P_{-+} &= N_{-+}/N, &&\text{corresponding to } \alpha\beta = -1,\\
P_{--} &= N_{--}/N, &&\text{corresponding to } \alpha\beta = 1,
\end{aligned}
$$

and the measured average of $\alpha\beta$ is

$$\langle \alpha\beta \rangle = \frac{N_{++} - N_{+-} - N_{-+} + N_{--}}{N}.$$

Each set of four coincidence counts corresponds to one particular setting of $\vec{A}$, $\vec{B}$, and yields a mean value $\langle \alpha\beta \rangle$. But in order to determine the correlation function $\langle \gamma \rangle$ featured in the BCHSH inequality, we need four mean values $\langle \alpha\beta \rangle$. Hence we choose, in succession, four different settings (Fig. 8.4c); four counting runs then yield four mean values

$$\langle \alpha_1\beta_1 \rangle, \quad \langle \alpha_1\beta_2 \rangle, \quad \langle \alpha_2\beta_1 \rangle, \quad \langle \alpha_2\beta_2 \rangle,$$

which determine the measured value of $\langle \gamma \rangle$ through

$$\langle \gamma \rangle = \langle \alpha_1\beta_1 \rangle + \langle \alpha_1\beta_2 \rangle + \langle \alpha_2\beta_1 \rangle - \langle \alpha_2\beta_2 \rangle.$$

8.9 The results of the first experiment at Orsay

These results are shown in Fig. 8.8. The angle θ which specifies the setting of the polarizers (Fig. 8.4c) is plotted horizontally, and the mean value $\langle \gamma \rangle$ vertically.

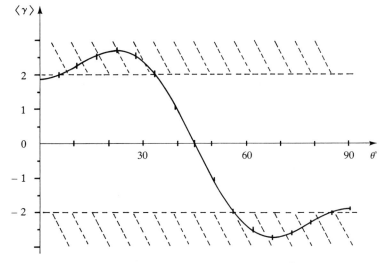

Fig. 8.8 The results of the first Orsay experiment.

The correlation function predicted by quantum mechanics reads

$$\langle \gamma \rangle = 3 \cos 2\theta - \cos 6\theta,$$

drawn as the continuous curve on the graph.† According to the BCHSH inequality

$$-2 \leq \langle \gamma \rangle \leq 2;$$

hidden-variable theories exclude the cross-hatched regions of the plane, which correspond to $\langle \gamma \rangle > 2$ and to $\langle \gamma \rangle < -2$.

The experimental results from 17 different values of θ are indicated on the figure by small vertical bars, each representing a value of $\langle \gamma \rangle$ plus or minus one standard deviation.

Clearly there can be no doubt that the BCHSH inequality is violated: eight of the experimental points fall outside the interval $[-2, 2]$. At the point where the violation is maximal ($\theta = 22.5°$), one finds

$$\langle \gamma \rangle = 2.70 \pm 0.015,$$

which represents a departure of over forty standard deviations from the extreme value 2. What is even more convincing is the precision with which the experimental points lie on the curve predicted by quantum mechanics. Quite evidently, for the EPR scenario one must conclude not only that hidden-variable theories fail, but that quantum mechanics is positively the right theory for describing the observations.

8.10 The relativistic test

The EPR experiment just described shows that the measurements in A and B are correlated. What is the origin of these correlations?

According to quantum theory, before the measurement each particle pair constitutes a single system extending from A to B, whose two parts are non-separable and correlated. This interpretation corresponds to a violation of Bell's inequality.

According to hidden-variable theories, the particle pair is characterized, at the instant of decay, by its hidden variable λ, which determines the correlation between the polarizations measured in A and B. This interpretation satisfies Bell's inequality.

Accordingly, the Orsay experiment supports the quantum interpretation (in

† The curve has been corrected for instrumental effects like solid angle, which explains why its ends are not precisely at 2 and −2.

terms of correlation between two parts A and B of a single system). However, to clinch this conclusion, one must ensure *that no influence is exerted* in the ordinary classical sense through some interaction propagated between the two detectors A and B, that is no influence which might take effect after the decay at S, and which might be responsible for the correlation actually observed. Let us therefore examine the Orsay apparatus in more detail (Fig. 8.9).

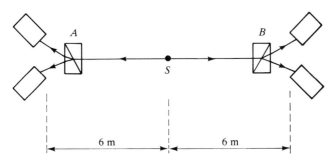

Fig. 8.9 Einsteinian non-separability.

When the detectors at A and B record a coincidence, this means that both have been triggered within a time interval of at most 10 ns, the resolving time of the circuit. Could it happen that, within this interval, A sends to B a signal capable of influencing the response of B? In the most favourable case such a signal would travel with the speed of light *in vacuo*, which according to relativity theory is the upper limit on the propagation speed of information, and thereby of energy. To cover the distance AB, which is 12 m (Fig. 8.9), such a signal would need 40 ns. This is too long by at least 30 ns, and rules out any causal links between A and B in the sense of classical physics. One says that *the interval between A and B is space-like*.

One of the advantages of the Orsay experiment is that it uses a very strong light-source, admitting sufficient distance between the detectors A and B while still preserving reasonable counting rates. By increasing the distance AB step by step, Aspect could check that the correlation persists, even when the interval between A and B becomes space-like. This is the check that guarantees that the two-photon system is non-separable irrespective of the distance AB.

It has become the custom to speak of *the principle of Einsteinian separability* in order to denote the absence of correlations between two events separated by a space-like interval. This is the principle that the Orsay experiment invites us to reconsider, even though our minds, used to the world at the macroscopic level, find it difficult to conceive of two 'microscopic' photons 12 m apart as a single indivisible object.

8.11 The final stage of the experiment at Orsay (1983)

Though the results of the first Orsay experiment are unarguable and clear-cut, the conclusion they invite is so startling that one should not be surprised at the appearance of a last-ditch objection, which as it happens gave the experimenters a great deal of trouble. In the preceding section we discussed the possible role of interactions between A and B operating after the decay at S, and duly eliminated this objection. But one can also ask whether correlations might be introduced through an interaction operating *before* the decay. We could imagine that the decay itself is preconditioned by the setting of the detectors A and B, such influences taking effect through the exchange of signals between detectors and source. No such mechanism is known *a priori*, but we do know that, if there is one, then Einsteinian non-separability would cease to be a problem, because the mechanism could come into action long before the decay, removing any reason for expecting a minimum 30 ns delay. Though such a scenario is very unlikely, the objection is a serious one and must be taken into account; to get round it, the experimenter must be able to choose the orientation of the detectors A and B at random after the decay has happened at S. In more picturesque language we would say that the two photons must leave the source without knowing the orientations of the polarizers A and B. Briefly put, this means that it must be possible to change the detector orientations during the 20 ns transits over SA or SB.

The solution adopted at Orsay employs periodic switching every 10 ns. These changes are governed by two independent oscillators, one for channel A and one for channel B. The oscillators are stabilized, but however good the stabilization it cannot eliminate small random drifts that are different in the two channels, seeing that the oscillators are independent. This ensures that the changes of orientation are random even though the oscillations are periodic, provided the experiment lasts long enough (one to three hours).

The key element of the second Orsay experiment is the *optical switch* (Fig. 8.10). In a water tank, a system of standing waves is produced by electro-acoustic excitation at a frequency of 25 MHz.†

The fluid keeps changing from a state of perfect rest to one of maximum agitation and back again. In the state of rest, the light-beam is simply transmitted. In the state of maximum agitation, the fluid arranges itself into a structure of parallel and equidistant plane layers, alternately stationary (nodal planes) or agitated (antinodal planes). Thus one sets up a lattice of net-like diffracting planes; the diffracted intensity is maximum at the Bragg angles, just

---

† 25 MHz corresponds to 10 ns between switchings.

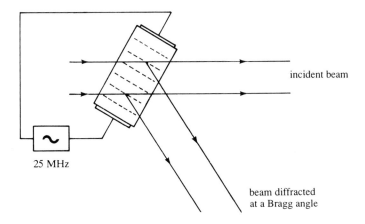

Fig. 8.10 The optical switch.

as in a crystal lattice. Here, the light-beam is deviated through 10^{-2} radians. The two numerical values, 25 MHz and 10^{-2} radians,† suffice to show the magnitude of the technical achievement. With the acoustic power at 1 watt, the system functions as an ideally efficient switch.

The second Orsay experiment (using the optical switches) is sketched in Fig. 8.11. In this set-up, the photons a and b leave S without 'knowing' whether they will go, the first to A_1 or A_2, and the second to B_1 or B_2.

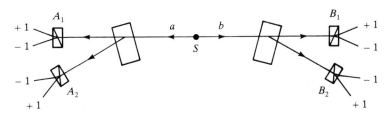

Fig. 8.11 The second Orsay experiment.

This second experiment is less precise than the first, because the light-beams must be very highly collimated in order to ensure efficient switching. Nevertheless its results exhibit an unambiguous violation of Bell's inequality, reaching five standard deviations at peak; moreover the results are entirely compatible with the predictions of quantum mechanics.

† An angle of 10^{-2} radians is less than $0.6°$; Fig. 8.10 shows it very much exaggerated in order to make the mechanism of the optical switch easier to visualize.

8.12 The principle of non-separability

Experiment has spoken. Half a century after the Como conference, Bohr's interpretation once again beats Einstein's, in a debate more subtle and also more searching. There were two conflicting theories:

Einstein	Bohr
hidden variables	quantum mechanics
realist	positivist
deterministic	probabilistic
separable	non-separable

The violation of the BCHSH inequality argues for Bohr's interpretation, all the more so as the measured values of $\langle \gamma \rangle$ are in close agreement with the predictions of quantum mechanics.

It remains to ask oneself just why hidden-variable theories do fail. Of the three basic assumptions adopted by such theories, namely realism, determinism, and separability, at least one must be abandoned. In the last resort, it is separability that seems to be the most vulnerable assumption. Indeed, one observes experimentally that the violation of the BCHSH inequality is independent of the distance between the two detectors A and B, even when this distance is 12 m or more.†

The principle of Einsteinian separability asserts that 'there are no correlations between two phenomena separated by a space-like interval'. In other words, no interaction can propagate faster than light *in vacuo*. In an EPR scenario this principle must be abandoned, and replaced by a principle asserting non-separability:

'In a quantum system evolving free of external perturbations, and from well-defined initial conditions, all parts of the system remain correlated, even when the interval between them is space-like.'

This assertion reflects properties of the state vector of a quantum system. For an EPR system, the state vector after the decay of the source reads

$$|\Phi\rangle = \frac{1}{\sqrt{2}} \left(|x_A, x_B\rangle + |y_A, y_B\rangle \right).$$

This expression combines the elements A and B in a non-separable manner, which is what explains the observed correlations. The truth is that all this has

---

† There are still die-hard advocates of determinism, who try to explain non-separability through non-local hidden variables. Such theories, awkward and barely predictive, are typically *ad hoc*, and fit only a limited number of phenomena. They are weakly placed to defend themselves against the interpretations furnished by quantum mechanics, which have the virtues of simplicity, elegance, efficiency, and generality, and which are invariably confirmed by experiment.

been well known ever since the beginnings of quantum mechanics, with the concept of the electron cloud as the most telling illustration. It is for instance hard to imagine separability between the 92 electrons of a uranium atom. What is new is that quantum mechanics, considered hitherto as a microscopic theory applicable on the atomic scale, is now seen to apply to a two-particle system macroscopically, on a scale of metres. The truly original achievement of Aspect's experiment is the demonstration of this fact.

Quantum objects have by no means exhausted their capacity to astonish us by their difference from the properties of the macroscopic objects in our everyday surroundings. In the preceding chapter we saw that a photon can interfere with itself. The present chapter has shown that two photons 12 m apart constitute but a single object. Thus it becomes ever more difficult to picture a photon through analogies with rifle bullets, surface waves in water, clouds in the sky, or with any other object of our familiar universe. Such partial analogies fail under attempts to make them more complete, and through their failure we discover new properties pertaining to quantum objects. The only fruitful procedure is to follow the advice of Niels Bohr, namely to bend one's mind to the new quantum concepts until they become habitual and thereby intuitive. Earlier generations of physicists have had to face similar problems. They had to progress from Aristotle's mechanics to Newton's, and then from Newton's to Einstein's. The same effort is now required of us, at a time favourable in that, by mastering the EPR paradox, quantum mechanics has just passed a particularly severe test with flying colours. From this point of view, the principle of non-separability seems as important as the principle of special relativity, and Aspect's experiment plays the same role now that the Michelson–Morley experiment played then.

8.13 Physicists and philosophers

Fifty years have elapsed since Einstein, Podolsky, and Rosen proposed their famous paradox: fifty years of intense activity by workers fascinated alike by physics and by philosophy.Today the EPR paradox is a paradox no longer: nature has declared unambiguously for quantum mechanics, and against any local hidden-variable theory.

It could be argued that all this effort has been quite useless. Seeing that quantum mechanics has worked perfectly well for more than half a century, why was it so necessary to put it back into the melting-pot, on the strength of nothing more than an *a priori* philosophical position of classical determinism? The present writer does not share this point of view. Physics benefits whenever one of its theories passes some very difficult test with great brilliance. For a start, we are the beneficiaries henceforth of the extension of the non-separability principle, which establishes quantum mechanics as an (also) macroscopic theory.

Moreover, at this point we can see the emergence of a new problem, illustrating yet again the fact that research in physics involves constant recommencement.

Classical Newtonian mechanics is a separable theory: witness its successful account of the Earth–Moon system in terms of two interacting but perfectly individualized bodies. Quantum mechanics by contrast is a non-separable theory, seeing that it describes the correlations between two atoms in terms of a single object whose dimensions, far from being confined to those of either atom, extend out to macroscopic distances. But we also know that Earth and Moon are conglomerations of atoms. How then can these atoms, non-separable quantum objects, generate separable classical objects merely by congregating in large numbers? How does quantum non-separability lead to the separability observed on the classical level? We have already met a problem of this kind in the last chapter. The motion of an electron through an electrostatic field is conveniently described by quantum mechanics, without any appeal to trajectories; the motion of an ensemble of electrons is conveniently described through relations between quantum-mechanical mean values, given by Ehrenfest's theorem, and these relations correspond exactly to Newton's second law of motion in classical mechanics. Thus one can calculate a mean trajectory for the electron beam, and this trajectory is the one observed in a cathode-ray tube.

This last example seems very suggestive of the path likely to be followed by developments in the future. It remains only to watch the evolution of physics until there emerges some brilliant young theorist, a new Ehrenfest, who can explain just how quantum non-separability turns into separability on the classical level. It will certainly not be easy, but quantum theorists are so accustomed to success that we can be reasonably optimistic.

Appendices to Chapter 8

1 Bell's inequality

1.1 *A theory that is deterministic and separable*

Suppose that the pair a, b emerging from S can be characterized by a hidden variable λ. The responses of the detectors A, B are $\alpha(A, \lambda)$ and $\beta(B, \lambda)$ respectively (Fig. 8.12). The theory is deterministic and separable:

- *deterministic*, because the results are determined by the hidden variable plus the settings of A and B;

• *separable*, because the response of A is independent of the response of B, and vice versa.

Since the value of λ is unknown and different for each pair, the responses of A and B seem random. Lacking information about λ, we characterize it by choosing a statistical distribution $\rho(\lambda)$. Thence we derive the distributions of the responses $\alpha(A, \lambda)$ and $\beta(B, \lambda)$, which can be compared with experiment.

Bell's inequalities have the great virtue that they apply to any hidden-variable theory, irrespective of our choice of $\rho(\lambda)$.

THEOREM 1. Consider four numbers α_1, α_2, β_1, and β_2, each of which can assume only the values 1 or -1. Then the combination

$$\gamma = \alpha_1\beta_1 + \alpha_1\beta_2 + \alpha_2\beta_1 - \alpha_2\beta_2$$

can assume only the values 2 or -2.

To prove the theorem we construct a truth table for all 16 possibilities, which shows that 2 and -2 are indeed the only possible values of γ.

THEOREM 2. Consider very many sets of four numbers $(\alpha_1, \alpha_2, \beta_1, \beta_2)$. The mean value of γ lies in the range $[-2, 2]$. In other words,

$$-2 \leq \langle \gamma \rangle \leq 2.$$

This is obvious, because every value of γ lies in this range, and so therefore must be the mean. (The end-points are included in order to allow for limiting cases.)

Note that both theorems are purely mathematical; neither involves any assumptions about physics.

1.2 *The BCHSH inequality*

Within the framework of a theory that is realist, deterministic, and separable, we can describe the photon pair in detail. Realism leads us to believe that polarization is an objective property of each member of the pair, independently of any measurements that may be made later. Determinism leads us to believe that the polarizations are uniquely determined by the decay cascade, and that they are fully specified by the hidden variable λ, which governs the correlation of the polarizations in A and B. Finally, separability leads us to believe that the measurements in A and B do not influence each other, which means in particular that the response of detector A is independent of the orientation of detector B.

Now consider a pair of photons a, b, characterized by a hidden variable λ. The responses of the apparatus in its four settings would be as follows:

$$\alpha_1 \text{ and } \beta_1 \text{ in the orientation } (A_1, B_1),$$
$$\alpha_2 \text{ and } \beta_2 \text{ in the orientation } (A_2, B_2),$$

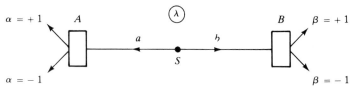

Fig. 8.12 Application of Bell's inequality to an EPR scenario.

α'_1 and β'_2 in the orientation (A_1, B_2),
α'_2 and β'_1 in the orientation (A_2, B_1),

(Recall that the variables α and β can take only the values 1 and -1.)

It is impossible in practice to make four measurements on one and the same pair of photons, because each photon is absorbed in the first measurement made on it; that is why we have spoken conditionally (i.e. of what the results *would* be). But if we believe that the photon correlations are governed by a theory that is realist, deterministic, and separable, then we are entitled to assume that the responses, of type α or type β, depend on properties that the photons possess before the measurement, so that the responses correspond to some objective reality. In such a framework we can appeal to the principle of separability, which implies, for instance, that detector A would give the same response to the orientations (A_1, B_1) and (A_1, B_2), because the response of A is independent of the orientation of B. Mathematically this is expressed by the relation

$$\alpha_1 = \alpha'_1.$$

Similarly one finds

$$\alpha_2 = \alpha'_2, \qquad \beta_1 = \beta'_1, \qquad \beta_2 = \beta'_2.$$

Thus we have shown that, for a given pair of photons, all possible responses of the apparatus in its four chosen settings can be specified by means of only four two-valued variables α_1, α_2, β_1, and β_2. This reduction from eight to four variables depends on the principle of separability. In this way we are led to a situation covered by Theorem 2, whence $-2 < \langle \gamma \rangle < 2$.

By making many measurements for each of the four settings we can determine the four mean values $\langle \alpha_1 \beta_1 \rangle$, $\langle \alpha_1 \beta_2 \rangle$, $\langle \alpha_2 \beta_1 \rangle$, $\langle \alpha_2 \beta_2 \rangle$, and thence the mean of the correlation function,

$$\langle \gamma \rangle = \langle \alpha_1 \beta_1 \rangle + \langle \alpha_1 \beta_2 \rangle + \langle \alpha_2 \beta_1 \rangle - \langle \alpha_2 \beta_2 \rangle.$$

2 The quantum calculation for an EPR scenario

The laboratory reference frame $Oxyz$ serves to specify the orientations of detectors and polarizers (Fig. 8.13).

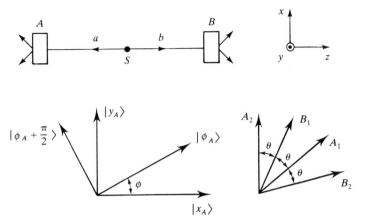

Fig. 8.13 Quantum calculation of EPR photon correlations for the settings chosen by Aspect.

Before any measurements have been made, the photon pair a,b forms a non-separable entity, represented by the vector

$$|\Phi\rangle = \frac{1}{\sqrt{2}}(|x_A, x_B\rangle + |y_A, y_B\rangle).$$

The act of measurement corresponds to passage to the ϕ-representation. Hence we require the transition amplitudes from the two states $|x_A, x_B\rangle$, $|y_A, y_B\rangle$ to the four states $|\phi_A, \phi_B\rangle$, $|\phi_A, \phi_B + \frac{1}{2}\pi\rangle$, $|\phi_A + \frac{1}{2}\pi, \phi_B\rangle$, $|\phi_A + \frac{1}{2}\pi, \phi_B + \frac{1}{2}\pi\rangle$.

The calculation, which follows the method described in Section 7.3, is rather lengthy [2]. The result on the other hand is quite simple, and reads

$$|\phi\rangle = \frac{1}{\sqrt{2}}[|\phi_A, \phi_B\rangle \cos(\phi_B - \phi_A)$$

$$-|\phi_A, \phi_B + \tfrac{1}{2}\pi\rangle \sin(\phi_B - \phi_A)$$

$$+|\phi_A + \tfrac{1}{2}\pi, \phi_B\rangle \sin(\phi_B - \phi_A)$$

$$+|\phi_A + \tfrac{1}{2}\pi, \phi_B + \tfrac{1}{2}\pi\rangle \cos(\phi_B - \phi_A)].$$

The square of each amplitude featured here represents the detection probability. For example, the probability of simultaneously detecting photon a polarized at an angle ϕ_A and photon b polarized at an angle ϕ_B is

$$\left[\frac{1}{\sqrt{2}} \cos(\phi_B - \phi_A)\right]^2 = \tfrac{1}{2} \cos^2(\phi_B - \phi_A).$$

By convention, we write the responses of detector A to a photon in state $|\phi_A\rangle$ (resp. $|\phi_A + \tfrac{1}{2}\pi\rangle$) as $\alpha = 1$ (resp. $\alpha = -1$); and similarly with β for detector B.

Let us analyse the four possible responses.

- $|\phi_A, \phi_B\rangle$ elicits $\alpha = 1$, $\beta = 1$, whence $\alpha\beta = 1$; the probability is

$$P_{++} = \tfrac{1}{2}\cos^2(\phi_B - \phi_A).$$

- $|\phi_A, \phi_B + \tfrac{1}{2}\pi\rangle$ elicits $\alpha = 1$, $\beta = -1$, whence $\alpha\beta = -1$; the probability is

$$P_{+-} = \tfrac{1}{2}\sin^2(\phi_B - \phi_A).$$

- $|\phi_A + \tfrac{1}{2}\pi, \phi_B\rangle$ elicits $\alpha = -1$, $\beta = 1$, whence $\alpha\beta = -1$; the probability is

$$P_{-+} = \tfrac{1}{2}sin^2(\phi_B - \phi_A).$$

- $|\phi_A + \tfrac{1}{2}\pi, \phi_B + \tfrac{1}{2}\pi\rangle$ elicits $\alpha = -1$, $\beta = -1$, whence $\alpha\beta = 1$; the probability is

$$P_{--} = \tfrac{1}{2}\cos^2(\phi_B - \phi_A).$$

The mean value of $\langle\alpha\beta\rangle_{AB}$ follows immediately as

$$\langle\alpha\beta\rangle_{AB} = P_{++} - P_{+-} - P_{-+} + P_{--}.$$

Hence

$$\langle\alpha\beta\rangle_{AB} = \cos^2(\phi_B - \phi_A) - \sin^2(\phi_B - \phi_A),$$

or in other words

$$\langle\alpha\beta\rangle_{AB} = \cos 2(\phi_B - \phi_A).$$

The setting chosen by Aspect is shown in Fig. 8.13. Corresponding to it we have the four terms

$$\langle\alpha_1\beta_1\rangle = \langle\alpha\beta\rangle_{A_1 B_1} = \cos 2(\phi_{B_1} - \phi_{A_1}) = \cos 2\theta,$$
$$\langle\alpha_1\beta_2\rangle = \langle\alpha\beta\rangle_{A_1 B_2} = \cos 2(\phi_{B_2} - \phi_{A_1}) = \cos 2\theta,$$
$$\langle\alpha_2\beta_1\rangle = \langle\alpha\beta\rangle_{A_2 B_1} = \cos 2(\phi_{B_1} - \phi_{A_2}) = \cos 2\theta,$$
$$\langle\alpha_2\beta_2\rangle = \langle\alpha\beta\rangle_{A_2 B_2} = \cos 2(\phi_{B_2} - \phi_{A_2}) = \cos 6\theta,$$

For comparison with Bell's inequality, we introduce the correlation function $\langle\gamma\rangle$:

$$\langle\gamma\rangle = \langle\alpha_1\beta_2\rangle + \langle\alpha_1\beta_2\rangle + \langle\alpha_2\beta_1\rangle - \langle\alpha_2\beta_2\rangle,$$

whence

$$\langle\gamma\rangle = 3\cos 2\theta - \cos 6\theta.$$

This correlation function, calculated according to quantum mechanics, is shown as the solid curve in Fig. 8.8.

Further reading

See Aspect (1976), Aspect, Dalibard, and Roger (1982), D'Espagnat (1979), and Hughes (1981) in the Bibliography.

Athene. The Greek goddess of wisdom. Known to Homer as Pallas Athene, and as Minerva to the Romans, in our day she is still the living symbol of science. (Alinari-Viollet)

9

Conclusions (how chance does yeoman's service)

When late I attempted your pity to move,
What made you so deaf to my prayers?
Perhaps it was right to dissemble your love,
But—why did you kick me downstairs?

Isaac Bickerstaffe

Bickerstaffe's lines reflect rather accurately the sufferings of a suitor frustrated by Nature in his pleas for classical determinism. It is evident that in the present state of physics all hopes of identifying science with determinism are in utter ruin. Here, at the end of this book, we are well placed to appreciate the reasons why.

It was only natural for physicists to be interested, at first, in the macroscopic world that surrounds us; in order to make quantitative predictions about it, they devised deterministic models which perform impeccably. Such are the origins of mechanics, of thermodynamics, of optics, of electromagnetism, and of relativity. These theories remain valid in the domains for which they were designed, and they continue in a state of vigorous development. But as regards fundamentals, physics today has its cutting edge on the microscopic level, where progress is achieved by means of probabilistic models, models that allow precise quantitative predictions for random phenomena.

Thus, chance is now well ensconced in physical theories applicable under a wide range of conditions: chance as a matter of ignorance in the statistical physics of Maxwell and Boltzmann; chance as a matter of conviction in the chaotic processes introduced by Poincaré; and chance unavoidable in quantum mechanics as formulated by Bohr and confirmed by Aspect's experiments at Orsay. In the first two cases, determinism remains unchallenged in principle but proves useless in practice; in the third case chance is inherent in the basic nature of microscopic processes, reducing determinism to a mere consequence of chance regarding mean values, that is on the macroscopic level.

Though the historic privileges of determinism have been revoked in favour of chance, there is absolutely no reason why we should allow this to distress us. In fields where deterministic models prove useless, the introduction of unfailingly operational probabilistic models allows prediction, which is the overriding objective of science. Phenomena that are inherently unpredictable at the level of single events become predictable, and with precision, as soon as the models are applied to statistical ensembles of such events. Physics undoubtedly relies on chance, but on chance confined, channelled, and mastered by the theory of probability. This theory brings the disorder apparent in the phenomena of nature into correspondence with a subtle order formulated through distribution laws, strange attractors, or vectors in Hilbert space. Is it not time therefore, at long last, to accept chance as fundamentally characteristic of Nature?

On this last point the physics community remains divided. By education, every physicist is imbued with an inheritance from classical, and deterministic, physics; and it is only natural that one should have a tendency to remain entrenched behind its paradigms (a jargon word coined as a didactic euphemism for received opinion). It would of course be very difficult to demonstrate conclusively that, as the case may be, Nature is basically deterministic, or basically random: philosophers have tried it and have failed. But, today, physics is progressing much faster by the use of random than of deterministic models: so much so that chance is bound to become more and more part of the language and eventually of the paradigms of physics. This book will have achieved its aims if it can help the reader to make the conceptual leap from common sense to the wisdom of physics, and from determinism to chance.

Bibliography

1. Adler, J. B. and Wainwright, T. E. (1959). Molecular motions. *Scientific American*, **201**, 113.
2. Aspect, A. (1976). Proposal for an experiment to test the nonseparability of quantum mechanics. *Physical Review*, **D14**, 1944.
3. Aspect, A., Dalibard, J., and Roger, G. (1982). Experimental test of Bell inequalities using time-varying analysers. *Physical Review Letters*, **49**, 1804.
4. Bergé, P., Pomeau, Y., Vidal, C., and Tuckermann, L. (1984). *Order within chaos*. Wiley, NY.
5. Cohen-Tannoudji, C., Diu, B., and Laloé, F. (1977). *Quantum mechanics* (2 vols). Wiley, NY.
6. Crutchfield, J. P., Farmer, J. D., Packard, N. H., and Shaw, R. S. (1986). Chaos. *Scientific American*, **255**, 38.
7. D'Espagnat, B. (1979). The quantum theory and reality. *Scientific American*, **241**, 128.
8. Ekeland, I. (1988). *Mathematics and the unexpected*. University of Chicago Press.
9. Feynman, R. P., Leighton, R. B., and Sands, M. (1963–65). *The Feynman lectures on physics* (3 vols). Addison-Wesley, Reading, MA.
10. Gamow, G. (1966). *Thirty years that shook physics*. Reprinted 1985, Dover, NY.
11. Grangier, P., Roger, G., and Aspect, A. (1986). 'Experimental evidence for a photon anticorrelation effect on a beam splitter: a new light on single-photon interference.' *Europhysics Letters*, **1**, 173.
12. Guinier, A. and Jullien, R. (1989). *The solid state* (tr. W J. Duffin). Oxford University Press.
13. Hughes, R. I. G. (1981). Quantum logic. *Scientific American*, **245**, 146.
14. Kittel, C. (1969). *Thermal physics*. Wiley, NY.
15. Lorenz, E. N. (1963). Deterministic non-periodic flow. *Journal of Atmospheric Sciences*, **20**, 130.
16. Mandelbrot, B. (1982). *The fractal geometry of nature*. Freeman, NY.
17. Marcus, P. M. and McFee, J. H. (1959). *Recent research in molecular beams* (ed. I. Eastermann). Academic Press, NY.
18. Messiah, A. (1961–62). *Quantum mechanics* (tr. G. M. Temmer and J. Potter; 2 vols). North-Holland, Amsterdam.
19. Moseley, H. G. (1913). *Philosophical Magazine*, **26**, 1024.
20. Moseley, H. G. (1914). *Philosophical Magazine*, **27**, 703.
21. Perrin, J. (1920). *Atoms* (tr. D. Ll. Hammick). London. Reprinted 1990, Ox Bow Press, Woodbridge CT.
22. Reif, F. (1967). *Statistical physics*. Vol. 5 in Berkeley Physics Course. McGraw-Hill, NY.
23. Robinson, A. L. (1986). Demonstrating single photon interference. *Science*, **231**, 671.
24. Rocard, Y. (1961). *Thermodynamics* (tr. C. R. S. Manders). London.

Index